Pseudo-Differential Operators
Theory and Applications
Vol. 1

Pseudo-Differential Operators: Theory and Applications is a series of moderately priced graduate-level textbooks and monographs appealing to students and experts alike. Pseudo-differential operators are understood in a very broad sense and include such topics as harmonic analysis, PDE, geometry, mathematical physics, microlocal analysis, time-frequency analysis, imaging and computations. Modern trends and novel applications in mathematics, natural sciences, medicine, scientific computing, and engineering are highlighted.

André Unterberger

Quantization and Arithmetic

Birkhäuser
Basel · Boston · Berlin

Author:

André Unterberger
Département Mathématiques et Informatique
Université de Reims
Moulin de la Housse
B.P. 1039
51687 Reims Cedex 2
France
e-mail: andre.unterberger@univ-reims.fr

2000 Mathematics Subject Classification: 11F11, 11F27, 47G30

Library of Congress Control Number: 2008927084

Bibliographic information published by Die Deutsche Bibliothek
Die Deutsche Bibliothek lists this publication in the Deutsche Nationalbibliografie;
detailed bibliographic data is available in the Internet at <http://dnb.ddb.de>.

ISBN 978-3-7643-8790-7 Birkhäuser Verlag AG, Basel · Boston · Berlin

© 2008 Birkhäuser Verlag AG
Basel · Boston · Berlin
P.O. Box 133, CH-4010 Basel, Switzerland
Part of Springer Science+Business Media
Printed on acid-free paper produced of chlorine-free pulp. TCF∞

ISBN 978-3-7643-8790-7   ISBN 978-3-7643-8791-4 (eBook)

9 8 7 6 5 4 3 2 1   www.birkhauser.ch

Contents

1 Introduction

Let $\chi^{(12)}$ be the unique *even* non-trivial Dirichlet character mod 12, and let $\chi^{(4)}$ be the unique (odd) non-trivial Dirichlet character mod 4. Consider on the line the distributions

$$\mathfrak{d}_{\text{even}}(x) = \sum_{m \in \mathbb{Z}} \chi^{(12)}(m) \, \delta \left(x - \frac{m}{\sqrt{12}} \right),$$

$$\mathfrak{d}_{\text{odd}}(x) = \sum_{m \in \mathbb{Z}} \chi^{(4)}(m) \, \delta \left(x - \frac{m}{2} \right). \tag{1.1}$$

Under a Fourier transformation, or under multiplication by the function $x \mapsto e^{i\pi x^2}$, the first (*resp.* second) of these distributions only undergoes multiplication by some 24th (*resp.* 8th) root of unity. Then, consider the metaplectic representation Met, a unitary representation in $L^2(\mathbb{R})$ of the metaplectic group $\widetilde{G}$, the twofold cover of the group $G = SL(2, \mathbb{R})$, the definition of which will be recalled in Section 2: it extends as a representation in the space $\mathcal{S}'(\mathbb{R})$ of tempered distributions. From what has just been said, if $\tilde{g}$ is a point of $\widetilde{G}$ lying above $g \in G$, and if $\mathfrak{d} = \mathfrak{d}_{\text{even}}$ or $\mathfrak{d}_{\text{odd}}$, the distribution $\mathfrak{d}^{\tilde{g}} = \text{Met}(\tilde{g}^{-1})\mathfrak{d}$ only depends on the class of g in the homogeneous space $\Gamma\backslash G = SL(2, \mathbb{Z})\backslash G$, up to multiplication by some phase factor, by which we mean any complex number of absolute value 1 depending only on $\tilde{g}$. On the other hand, a function $u \in \mathcal{S}(\mathbb{R})$ is perfectly characterized by its scalar products against the distributions $\mathfrak{d}^{\tilde{g}}$, since one has for some appropriate constants C_0, C_1 the identities

$$\int_{\Gamma\backslash G} |\langle \mathfrak{d}^{\tilde{g}}_{\text{even}}, u \rangle|^2 \, dg = C_0 \, \| u \|_{L^2(\mathbb{R})}^2 \qquad \text{if } u \text{ is even},$$

$$\int_{\Gamma\backslash G} |\langle \mathfrak{d}^{\tilde{g}}_{\text{odd}}, u \rangle|^2 \, dg = C_1 \, \| u \|_{L^2(\mathbb{R})}^2 \qquad \text{if } u \text{ is odd}. \tag{1.2}$$

Each of the two formulas is called a resolution of the identity since its polarized version makes it possible to write a general function u with a given parity as an integral superposition of the corresponding distributions $\mathfrak{d}^{\tilde{g}}$. In view of these equations, together with the fact that the distributions $\mathfrak{d}^{\tilde{g}}$ of a given parity are essentially permuted under the metaplectic representation, the family $(\mathfrak{d}^{\tilde{g}})$, $\tilde{g}$ denoting now a point in an appropriate homogeneous space of $\widetilde{G}$ covering $\Gamma\backslash G$, will be called a family of coherent states for the metaplectic representation: admittedly, this terminology will be frowned upon by some readers, since the distributions $\mathfrak{d}^{\tilde{g}}$ do not lie in $L^2(\mathbb{R})$.

In spite of this, one can still characterize a linear operator $A \colon \mathcal{S}'(\mathbb{R}) \to \mathcal{S}(\mathbb{R})$, preserving the parity of functions, by means of its matrix elements against the given family. By this, we mean the set of scalar products $(\mathfrak{d}^{\tilde{g}_2} \,|\, A \mathfrak{d}^{\tilde{g}_1})$: one can consider only the even (*resp.* odd) coherent states $\mathfrak{d}^{\tilde{g}}$ in the case when, moreover,

A kills all odd (*resp.* all even) functions, in which case we shall say that A is of even-even (*resp.* odd-odd) type. A very different question, central to the present work, is to what extent an operator A can be recovered from its diagonal matrix elements, *i.e.*, those for which $g_1 = g_2$.

Readers familiar with the so-called Wick symbol of an operator (a notion originating from Physics) or with the Berezin theory of quantization, will have a feeling of *déjà vu*. Indeed, instead of starting from the arithmetic distribution $\mathfrak{d}_{\mathrm{even}}$, let us start from the standard Gaussian function $x \mapsto 2^{\frac{1}{4}} e^{-\pi x^2}$: it is invariant, up to phase factors, under the operators of the metaplectic representation lying above the maximal compact subgroup $K = SO(2)$ of G, which makes it possible to build a family (u_{gK}) of coherent states (belonging this time to the space $L^2(\mathbb{R})$) parametrized by points in G/K. The associated family of diagonal matrix elements $(u_{gK} \,|\, A\, u_{gK})$ is just a symbol of some Berezin kind of the operator A: similar considerations bring to light the Wick symbol, but using the Heisenberg representation in place of the metaplectic representation. We identify in the usual way the homogeneous space G/K with the hyperbolic upper half-plane Π: by the way, the space $\Gamma\backslash G$, which is up to finite covering the parametrizing space of the family of arithmetic coherent states introduced above, can be identified, just as classically, with a set of lattices in the plane. It is one of the purposes of the present work to push the analogy between Γ and K as far as possible, staying entirely within classical analysis, *i.e.*, away from adeles: there is considerable room in $\mathbb{R}^2$ as soon as distributions enter the picture, and even though there is no fundamental domain for the linear action of Γ in $\mathbb{R}^2$, there is a well-defined concept of automorphic distribution in the plane [30, 31]. This notion is advantageous on several accounts: in particular, after it has been transferred in the right way (to wit, under some associate of the Radon transformation) from the half-plane to the plane, the operator $\Delta - \frac{1}{4}$ factors as $\pi^2 \mathcal{E}^2$, where $2i\pi\mathcal{E} = x\frac{\partial}{\partial x} + \xi\frac{\partial}{\partial \xi} + 1$ is the Euler operator in the plane. As a consequence, spectral decompositions of automorphic functions give way, in the plane, to decompositions of automorphic distributions into homogeneous components: this makes explicit calculations easier in general.

Coming back to the family of matrix elements $(u_z \,|\, A u_z)_{L^2(\mathbb{R})}$, where $z \in \Pi$ and A is assumed to be of even-even type, understanding the amount of information carried by this function demands that we should first characterize A by a symbol in some *good* symbolic calculus: in this case, it is of course the Weyl calculus, for many reasons, one of which is that it establishes an isometry Op from the space $L^2(\mathbb{R}^2)$ to the space of Hilbert-Schmidt operators in $L^2(\mathbb{R})$. The operator $A = \mathrm{Op}(h)$ is of even-even type if and only if its symbol h is an even function invariant under a rescaled version $\mathcal{G}$ (by a factor 2) of the symplectic Fourier transformation in the plane. If such is the case, and if one sets $(\mathcal{C}h)(z) = (u_z \,|\, A u_z)_{L^2(\mathbb{R})}$, one has the spectral-theoretic equation

$$\| \mathcal{C}h \|_{L^2(\Pi)} = 2 \, \| \, \Gamma(i\pi\mathcal{E}) \, h \, \|_{L^2(\mathbb{R}^2)} \,, \tag{1.3}$$

which captures exactly the loss of information incurred in considering $\mathcal{C}h$ instead of h, or A. Since the Gamma function is rapidly decreasing at infinity on vertical lines in the complex plane, the operator $\mathcal{C}$ is far from invertible: on the other hand, the Gamma function does not vanish at any finite point, which implies that the map $\mathcal{C}$ is one-to-one, even when given the space of all even-even tempered distributions as a domain. This last fact could also be obtained by an argument of analytic continuation since, as will be recalled, u_z is the product of $(\mathrm{Im}\,(-z^{-1}))^{\frac{1}{4}}$ by an antiholomorphic function of z: then, the set of matrix elements $(u_{z_2}\,|\,Au_{z_1})$ is characterized by its subset of diagonal ones.

Let us now switch to the arithmetic situation, assuming that $h \in \mathcal{S}(\mathbb{R}^2)$ so that the associated Weyl operator should act from $\mathcal{S}'(\mathbb{R})$ to $\mathcal{S}(\mathbb{R})$, that A is of even-even type and that $h(0) = 0$: we also assume that h is a radial function – most of the analysis remains without this condition, but must be formulated in terms of automorphic distributions, not automorphic functions – so that the scalar product $(\mathfrak{d}_{\mathrm{even}}^{\tilde{g}}\,|\,A\,\mathfrak{d}_{\mathrm{even}}^{\tilde{g}})$ depends only on $z = g.i \in \Pi$. The function $(\mathcal{A}h)(z)$ so defined is automorphic, and one has the identity, again of a spectral-theoretic type,

$$\|\,\mathcal{A}h\,\|_{L^2(\Gamma\backslash\Pi)} = \left(\frac{8}{\pi}\right)^{\frac{1}{2}}\,\|\,(1 - 2^{2i\pi\mathcal{E}})(1 - 3^{2i\pi\mathcal{E}})\,\zeta(1 - i\pi\mathcal{E})\,h\,\|_{L^2(\mathbb{R}^2)}.\qquad(1.4)$$

Since the zeta function has no zero on the line $\mathrm{Re}\,s = 1$, the inverse operator $\mathcal{A}^{-1}$ becomes continuous if followed by the spectral projection, relative to the self-adjoint operator $2\pi\mathcal{E}$, corresponding to any closed interval not containing any point $\frac{2\pi n}{\log 2}$ or $\frac{2\pi n}{\log 3}$ with $n \in \mathbb{Z}$. Of course, we cannot say that a function h, homogeneous of degree $-1 - i\lambda$ for such an exceptional value of λ, lies in the nullspace of $\mathcal{A}$ since we have to work with symbols in $\mathcal{S}(\mathbb{R}^2)$. Needless to say, something analogous works with odd functions too. Finally, the reader may wish to know what will happen if one uses the Dirac comb, the sum of unit masses at points of $\mathbb{Z}$, in place of $\mathfrak{d}_{\mathrm{even}}$ or $\mathfrak{d}_{\mathrm{odd}}$. Then, one cannot use the full modular group Γ, but only a certain subgroup, isomorphic to the so-called Hecke congruence group $\Gamma_0(2)$: the fundamental domain, in this case, has two cusps, which complicates a little bit the discussion, but not much. The matters discussed so far make up the first chapter of the present work.

It is interesting to compare (1.3) and (1.4), especially in view of the fact that, as will be discussed presently, there is a very natural generalization of the first identity in which the operator $\Gamma(i\pi\mathcal{E})$ has to be replaced by $\Gamma(\tau + \frac{1}{2} + i\pi\mathcal{E})$ for some real number $\tau > -1$ but otherwise arbitrary. One might expect that it should be possible, generalizing the second identity in a similar way, to manage so as to let the restriction of the function zeta to some possibly arbitrary, or at least not overspecialized, line enter the picture: then, the non-existence of zeros of zeta on such a line would be given an interesting interpretation. For some reasons which seem to us rather deep, and which leave room for further investigation, we were disappointed in this hope, but not completely. Before we come to this point, let us describe the generalization of (1.3), depending on the parameter τ, which we have in mind.

The idea, certainly not a novel one – but only up to some point – consists in regarding the even and odd parts of the metaplectic representation, up to unitary equivalence, as special cases of the holomorphic discrete series of $SL(2, \mathbb{R})$ or, more precisely, of a prolongation of the holomorphic discrete series of the universal cover of that group: the parameter τ labelling the representations in this series lies in the interval $]-1, \infty[$, the two special cases already considered corresponding to the values $\tau = \mp\frac{1}{2}$. The representation obtained for each value of τ has at least two useful realizations: one, denoted as $\mathcal{D}_{\tau+1}$, in a weighted L^2-space $H_{\tau+1}$ of functions on the half-line $(0, \infty)$, and one, denoted as $\pi_{\tau+1}$, in a space of holomorphic functions in Π, easily described only in the case when $\tau \geq 0$. In the first realization, there is a family $(\psi_z^{\tau+1})_{z \in \Pi}$ of functions on the half-line substituting for the former family (u_z): the family of diagonal matrix elements of an operator in $H_{\tau+1}$ against the set of coherent states just referred to is called the Berezin-covariant symbol of A.

Chapter 2 is devoted to the construction of a *good* symbolic calculus of operators in $H_{\tau+1}$. Let us hasten to say that this calculus is none of the calculi, using Π as a phase space (this is the space where symbols live), some readers may be familiar with, such as the Berezin calculus [2, 3], or the active-passive calculus [26, 29]. It will be necessary, in Section 6, to recall a few facts concerning these calculi: but this will only serve as a preparation to the introduction, in Section 7, of the horocyclic calculus. The construction of this second-generation calculus is more involved than that of the preceding ones, as it does not admit any obvious definition. The phase space is in this case $\mathbb{R}^2$, and symbols have to satisfy some specific symmetry property, expressed with the help of the rescaled Fourier transformation $\mathcal{G}$. This is the good calculus we had been aiming for, and (1.3) generalizes in the way already alluded to.

The third chapter revisits the notion of arithmetic coherent states in connection with the family of representations $(\mathcal{D}_{\tau+1})$, taking advantage of the symbolic calculi studied in Chapter 2. We consider the diagonal matrix elements of operators in $H_{\tau+1}$, characterized by their horocyclic symbols, against a τ-dependent family $(\mathfrak{s}_\tau^{\tilde{g}})_{g \in G}$ of arithmetic coherent states: these are built from a discrete measure $\mathfrak{s}_\tau$ on the half-line, invariant, up to phase factors, under all transformations $\mathcal{D}_{\tau+1}(\tilde{g})$, $\tilde{g}$ being an element of the universal cover of $SL(2, \mathbb{R})$ lying above a matrix $g \in SL(2, \mathbb{Z})$. Even though such a discrete measure is not unique, there is always at least one possible choice, obtained with the help of some power of the Ramanujan Δ-function. The resolution of the identity expressed in (1.2) easily generalizes in the obvious way. It is much more difficult, and this is the object of Chapter 3, to extend (1.4) to the τ-dependent case. The horocyclic calculus could not be dispensed with at this point, but there are other difficulties as well, which demand reconsidering in particular the Rankin-Selberg unfolding method of modular form theory. The version needed here starts from the consideration of series of the Poincaré style built not from the function $z = x + iy \mapsto y^s$, in the way Eisenstein's is, but from the Whittaker function $z \mapsto y^{\frac{1}{2}} K_{s-\frac{1}{2}}(2\pi ky) e^{2i\pi kx}$. Yes, the series would diverge, but one can bypass this difficulty in a certain canonical way, which makes it possible to complete the proof of the the τ-dependent generalization of (1.4).

This is the spectral decomposition of the automorphic function taking the place of the function $\mathcal{A}h$ in (1.4): instead of the zeta function, it is a certain convolution L-function, to wit $L(\overline{f} \otimes f, s)$, built with the help of the modular form f of real weight $\tau + 1$ used in the construction of $\mathfrak{s}_\tau$, that appears. This function has to be considered on the "spectral line" Re $s = \frac{1}{2}$. When $\tau = -\frac{1}{2}$ (the same goes when $\tau = \frac{1}{2}$), it is essentially (up to one or two elementary extra factors) the restriction of zeta to the boundary of the critical strip. On the contrary, when $\tau + 1$ is an even integer and f is a Hecke cusp-form, it follows from results of Shimura [23] (which we learned from [13, 14]) that the function $L(\overline{f} \otimes f, s)$ is "divisible" by $\zeta(s)$, thus letting the critical zeros of zeta participate in the non-invertibility of the map $\mathcal{A}$. It would certainly be a nice thing if a τ-dependent theory were available: it is rather unlikely – as long as the Riemann hypothesis has not been proved – that a spectral-theoretic interpretation of all non-trivial zeros of zeta (the "Hilbert-Polya dream" [17, p. 7]) could exist; having such a theory for those (hopefully non-existent) lying on any given line Re $s = a$, $\frac{1}{2} < a < 1$, is another matter. However, the move – in the non-arithmetic situation – from $\Gamma(i\pi\mathcal{E})$ to $\Gamma(\tau + \frac{1}{2} + i\pi\mathcal{E})$ does not generalize in the way one might have hoped for to the arithmetic case. On the other hand, since much depends on some better understanding of holomorphic modular forms of weight $\tau + 1$ and their associated convolution L-functions, results in the desired direction are not yet to be excluded, at least for some very special values of τ: but nothing easy can be expected.

Going beyond the case of radial (horocyclic) symbols, as treated in Section 10, is not completely obvious, as it is tantamount to substituting the homogeneous space $SL(2,\mathbb{Z})\backslash SL(2,\mathbb{R})$ for the double quotient $SL(2,\mathbb{Z})\backslash SL(2,\mathbb{R})/SO(2)$, and it is the subject of Section 11. Avoiding the use of the three-dimensional first space is nevertheless possible, relying on the concept of automorphic distributions ($SL(2,\mathbb{Z})$-invariant distributions in the plane) already referred to in this introduction and used for a different (in some sense dual) purpose in automorphic pseudodifferential analysis [31]: there, non-holomorphic modular distributions were used as symbols while, in the present work, it is the functions – or, rather, distributions – the operators are applied to that carry the arithmetic.

In the fourth, and last, chapter, we come back to the Weyl calculus, and obtain in a natural way a generalization of (1.4) in which the product of two factors on the right-hand side is replaced by an arbitrary partial product of the Eulerian expansion of the operator $(\zeta(2i\pi\mathcal{E}))^{-1}$. It requires that one substitute for the distribution $\mathfrak{d}_{\mathrm{even}}$ in (1.1) a finite collection of distributions, depending on some integer N, the product of 4 by a squarefree odd integer but otherwise arbitrary. Again (Theorem 12.4), a formula of "resolution of the identity" exists: it is especially interesting to analyze what happens when $N \to \infty$, more precisely when the set of prime divisors of N tends to the set of all prime numbers. In connection with a study of the distributions in the plane obtained in the process, we make the first few steps, in the last section, towards the development of an analysis specifically adapted to the study of *combs* or of their associated Dirichlet series. Note that it

Chapter 1

Weyl Calculus and Arithmetic

In this chapter, we consider the even and odd parts of the metaplectic representation of $\widetilde{G}$ (the twofold cover of $G = SL(2,\mathbb{R})$) in $L^2(\mathbb{R})$. Two families of coherent states, parametrized by the hyperbolic half-plane Π, and denoted as (u_z) and (u_z^1), are built with the help of the eigenstates with lowest energy levels of the harmonic oscillator. Equation (1.3) will be considered in Section 2: this will provide an opportunity to recall some of the main properties of the Weyl calculus and related concepts; the Weyl calculus is also a model for other symbolic calculi to be introduced later. Families of coherent states of an arithmetic nature will be constructed in Section 3: they are related to the Dedekind eta-function, a 24th root of the Ramanujan Δ-function. A succession of two intertwining operators, one a quadratic change of variable from the line to the half-line, the other a version of the Laplace transformation, will provide the link: the latter one will also be used later, in a τ-dependent context. Section 4 is devoted to calculations regarding the Wigner function of a pair of arithmetic coherent states. The spectral resolution of the automorphic function obtained from the diagonal matrix elements of an operator with a radial Weyl symbol against a family of coherent states of an arithmetic nature is finally obtained in Section 5.

2 A non-arithmetic prologue

This section has a preparatory status: its role is to familiarize the reader with some of the notions and methods, such as the one-dimensional metaplectic representation and Weyl calculus, generalizations of which will play a role throughout the book. We have not found it necessary at the present stage to give a complete proof of Theorem 2.1 below – which will be superseded by Theorem 7.7 – rather to explain its significance.

Let $\widetilde{G}$, known as the metaplectic group, be the twofold cover of the group $G = SL(2,\mathbb{R})$: recall [35] that there exists a certain unitary representation Met of

$\widetilde{G}$ in $L^2(\mathbb{R})$, called the metaplectic representation. It preserves the Schwartz space $\mathcal{S}(\mathbb{R})$ (the space of C^∞ vectors of Heisenberg's representation) and extends as a representation within the dual space $\mathcal{S}'(\mathbb{R})$. The set of all unitary operators $\mathrm{Met}(\tilde{g})$, $\tilde{g} \in \widetilde{G}$, is generated as a group by the operators of the following three species: (i) transformations $u \mapsto v$, $v(x) = a^{-\frac{1}{2}} u(a^{-1}x)$, $a > 0$; (ii) multiplications by exponentials $\exp i\pi c x^2$, c real; (iii) $e^{-\frac{i\pi}{4}}$ times the Fourier transformation $\mathcal{F}$, normalized as

$$(\mathcal{F}u)(\xi) = \int_{-\infty}^{\infty} u(x)\, e^{-2i\pi x\xi}\, dx. \tag{2.1}$$

These three transformations are associated with points $\tilde{g}$ that lie above the points $g = \left(\begin{smallmatrix} a & 0 \\ 0 & a^{-1} \end{smallmatrix}\right)$, $\left(\begin{smallmatrix} 1 & 0 \\ c & 1 \end{smallmatrix}\right)$ and $\left(\begin{smallmatrix} 0 & 1 \\ -1 & 0 \end{smallmatrix}\right)$ of G. As a consequence, Met is not an irreducible transformation, but acts within $L^2_{\mathrm{even}}(\mathbb{R})$ and $L^2_{\mathrm{odd}}(\mathbb{R})$ separately: the two terms can then be shown to be acted upon in an irreducible way.

The Weyl pseudodifferential calculus is the rule [36] that associates a linear operator $\mathrm{Op}(h)$ acting on functions of *one variable* to functions h of *two variables* according to the following defining formula:

$$(\mathrm{Op}(h)u)(x) = \int h\left(\frac{x+y}{2}, \eta\right) u(y)\, e^{2i\pi(x-y)\eta}\, dy\, d\eta, \qquad u \in \mathcal{S}(\mathbb{R}). \tag{2.2}$$

The (unique) function h is called the *symbol* of the operator $\mathrm{Op}(h)$, and the rule (or linear map) Op is the first example of *symbolic calculus* (a.k.a. pseudodifferential calculus, or analysis) we shall encounter in this work: we shall introduce other species in Chapter 2.

When $h \in \mathcal{S}(\mathbb{R}^2)$, the operator $\mathrm{Op}(h)$ sends the whole distribution space $\mathcal{S}'(\mathbb{R}^2)$ to $\mathcal{S}(\mathbb{R}^2)$; when $h \in \mathcal{S}'(\mathbb{R}^2)$, it is still defined as an operator from $\mathcal{S}(\mathbb{R}^2)$ to $\mathcal{S}'(\mathbb{R}^2)$. Between these two extreme situations lies the fact that Op establishes an isometry from the space $L^2(\mathbb{R}^2)$ to the Hilbert space of Hilbert-Schmidt operators on $L^2(\mathbb{R})$.

Before proceeding any further, note that it is traditional, in pseudodifferential analysis, to denote the current point of the *phase space* $\mathbb{R}^2$ as (x, ξ) or sometimes (y, η): unfortunately, $z = x + iy$ is also the current point of the upper half-plane, which will force us, at times, to denote the current point of $\mathbb{R}^2$ as (q, p), a notation inspired from quantum mechanics.

The operator $\mathrm{Op}(h)$ is called the operator with symbol h. On the other hand, one introduces, for any pair u, v of functions in $\mathcal{S}(\mathbb{R})$, the *Wigner function* $W(v, u)$ on $\mathbb{R}^2$ defined as

$$W(v, u)(x, \xi) = 2 \int_{-\infty}^{\infty} \bar{v}(x+t)\, u(x-t)\, e^{4i\pi t\xi}\, dt. \tag{2.3}$$

Then, as is easily seen, $W(v, u) \in \mathcal{S}(\mathbb{R}^2)$, and for every $h \in \mathcal{S}'(\mathbb{R}^2)$ the formula

$$(v|\mathrm{Op}(h)u) = \langle h, W(v, u)\rangle = \int_{\mathbb{R}^2} h(x, \xi)\, W(v, u)(x, \xi)\, dx\, d\xi \tag{2.4}$$

holds: observe that we define scalar products $(\,|\,)$ as being antilinear with respect to the argument *on the left*, whereas straight brackets $\langle\,,\,\rangle$ denote bilinear operations. The Wigner function $W(v,u)$ is also the symbol of the rank-one operator $w \mapsto (v|w)u$. It makes sense as an element of $\mathcal{S}'(\mathbb{R}^2)$ as soon as u, v both lie in $\mathcal{S}'(\mathbb{R})$.

Warning: in [31, p. 11], the 2 in front of the right-hand side of (2.3) was unfortunately omitted. However, all computations in Sections 1–17 were made with the correct Wigner function, including the factor 2; the calculations in Section 18 were based on the wrong formula, so that all computations in that section based on the use of the Wigner function led to results which ought to be multiplied by 2.

The Weyl calculus satisfies a certain *covariance rule*, which expresses its coherence with the metaplectic representation on one hand, the linear action of G on the phase space on the other: for every $\tilde{g} \in \tilde{G}$ lying above some point $g \in G$, and every tempered distribution h on $\mathbb{R}^2$, one has

$$\mathrm{Met}(\tilde{g})\,\mathrm{Op}(h)\,\mathrm{Met}(\tilde{g})^{-1} = \mathrm{Op}(h \circ g^{-1}). \tag{2.5}$$

On the phase space $\mathbb{R}^2$, we introduce the version of Euler's operator defined as

$$2i\pi\,\mathcal{E} = x\,\frac{\partial}{\partial x} + \xi\,\frac{\partial}{\partial \xi} + 1. \tag{2.6}$$

The operator $\mathcal{E}$, with initial domain $\mathcal{S}(\mathbb{R}^2)$, or even the space of C^∞ functions with a compact support not containing 0, is essentially self-adjoint in $L^2(\mathbb{R})$ and its spectrum is the real line: generalized eigenfunctions of this operator are thus just homogeneous functions of degree $-1-i\lambda$, such a function corresponding to the eigenvalue $-i\lambda$ of the operator $2i\pi\mathcal{E}$. The spectral decomposition with respect to this operator, in other words the decomposition of a function as an integral of homogeneous components of degrees $-1 - i\lambda$, $\lambda \in \mathbb{R}$, makes it possible to compute, often in an explicit way, functions, in the spectral-theoretic sense, of $\mathcal{E}$: several such formulas are obtained by integral superposition, starting from the fact that for every $t > 0$, one has

$$(t^{2i\pi\mathcal{E}}\,h)(x,\,\xi) = t\,h(tx,\,t\xi). \tag{2.7}$$

We shall use the rescaled version $\mathcal{G}$ of the symplectic Fourier transformation (this is the Fourier transformation in which one sign in the exponent has been changed: then, it becomes invariant under linear changes of coordinates associated to matrices in $SL(2,\mathbb{R})$, not only those in $SO(2)$) defined as $\mathcal{G} = 2^{i\pi\mathcal{E}}\,\mathcal{F}\,2^{-i\pi\mathcal{E}} = 2^{2i\pi\mathcal{E}}\,\mathcal{F}$, i.e.,

$$(\mathcal{G}\,h)(x,\,\xi) = 2\int_{\mathbb{R}^2} h(y,\,\eta)\,e^{4i\pi\,(x\eta - y\xi)}\,dy\,d\eta. \tag{2.8}$$

In the Weyl calculus, it plays the following role: with $\check{u}(t) = u(-t)$, one has

$$\mathrm{Op}(\mathcal{G}\,h)\,u = \mathrm{Op}(h)\,\check{u} \tag{2.9}$$

for every pair $(h, u) \in \mathcal{S}'(\mathbb{R}^2) \times \mathcal{S}(\mathbb{R})$ (or $\mathcal{S}(\mathbb{R}^2) \times \mathcal{S}'(\mathbb{R})$). Since (a very special case of the covariance property) an operator $\mathrm{Op}(h)$ commutes with the transformation $u \mapsto \check{u}$ if and only if its symbol satisfies the identity $h(x, \xi) = h(-x, -\xi)$, one may introduce the self-explaining notions of even-even symbols (the even and $\mathcal{G}$-invariant symbols), odd-odd symbols (the even symbols transforming to their negatives under $\mathcal{G}$), etc. ... Let us insist that the involution $\mathcal{G}$ will play an essential role throughout this work: it commutes with the linear changes of coordinates of $\mathbb{R}^2$ associated to matrices in G.

Let u_i (the subscript i corresponds to the base-point of the upper half-plane) be the $L^2(\mathbb{R})$-normalized standard Gaussian function such that $u_i(t) = 2^{\frac{1}{4}} e^{-\pi t^2}$. It is easy to check that

$$W(u_i, u_i)(x, \xi) = 2\, e^{-2\pi (x^2 + \xi^2)}. \tag{2.10}$$

Also, set $u_i^1(t) = 2^{\frac{5}{4}} \pi^{\frac{1}{2}} t\, e^{-\pi t^2}$, so that

$$W(u_i^1, u_i^1)(x, \xi) = 2\,[4\pi (x^2 + \xi^2) - 1]\, e^{-2\pi (x^2 + \xi^2)}. \tag{2.11}$$

If $\tilde{g} \in \widetilde{G}$ lies above $g \in SO(2)$, one can verify that, under the transformation $\mathrm{Met}(\tilde{g})$, each of the two functions u_i and u_i^1 is multiplied by some complex number of modulus 1: which number is to be found in [30, p. 120–121] or [31, p. 16]. Consequently, if $g = \begin{pmatrix} a & b \\ c & d \end{pmatrix}$ and $z = g.i = \frac{ai+b}{ci+d}$ is an arbitrary element of the upper half-plane Π, finally if $\tilde{g}$ is any of the two elements of the metaplectic group lying above g, the functions $\mathrm{Met}(\tilde{g})\, u_i$ and $\mathrm{Met}(\tilde{g})\, u_i^1$ only depend on z, up to the multiplication by some ($\tilde{g}$-dependent) complex number of absolute value 1: actually, the first function is a multiple (by some phase factor) of the function u_z, and the second one is a multiple of u_z^1, with

$$u_z(t) = 2^{\frac{1}{4}} \left(\mathrm{Im} \left(-\frac{1}{z} \right) \right)^{\frac{1}{4}} \exp \left(i\pi\, \frac{t^2}{z} \right),$$

$$u_z^1(t) = 2^{\frac{5}{4}} \pi^{\frac{1}{2}} \left(\mathrm{Im} \left(-\frac{1}{z} \right) \right)^{\frac{3}{4}} t \exp \left(i\pi\, \frac{t^2}{z} \right). \tag{2.12}$$

One has the equation (in which $d\mu(z) = y^{-2}\, dx\, dy$) [31, p. 16]

$$\int_\Pi |(u_z^1 \,|\, u)|^2\, d\mu(z) = 8\pi \, \| u \|_{L^2(\mathbb{R})}^2 \tag{2.13}$$

for every square-integrable odd function on the line, but nothing similar holds in the space of even functions (u_z^1 being replaced by u_z) because the corresponding summand of the metaplectic representation is not square integrable. Nevertheless, in view of the irreducibility of the metaplectic representation when restricted to functions in $L^2(\mathbb{R})$ of a definite parity, the set $\{u_z : z \in \Pi\}$ (resp. $\{u_z^1 : z \in \Pi\}$) is total in $L^2_{\mathrm{even}}(\mathbb{R})$ (resp. $L^2_{\mathrm{odd}}(\mathbb{R})$). As a consequence, an operator $\mathrm{Op}(h)$ with $h \in$

$\mathcal{S}'_{\text{even}}(\mathbb{R}^2)$, is characterized by the pair of functions $(z_1, z_2) \mapsto (u_{z_2} \,|\, \text{Op}(h)u_{z_1})$ and $z \mapsto (u^1_{z_2} \,|\, \text{Op}(h)u^1_{z_1})$: recall from the introduction that these functions are the matrix elements of the operator $\text{Op}(h)$ against the family of coherent states $\{u_z \colon z \in \Pi\}$ (*resp.* $\{u^1_z \colon z \in \Pi\}$). The even-even (*resp.* odd-odd) part of the symbol is characterized by the first (*resp.* the second) of these two functions. Note that the first function (nothing would have to be changed if concerned with the second one) is left invariant if u_{z_1} is changed to $\exp(i\theta)\,u_{z_1}$ while u_{z_2} is multiplied by the same phase factor $\exp(i\theta)$: this is why, up to some extent, we only need to know the metaplectic representation as a *projective* representation, *i.e.*, a representation "up to unitary factors"; then, we may as well consider it as a representation of G rather than $\widetilde{G}$.

Theorem 2.1. *Given* $h \in \mathcal{S}'(\mathbb{R}^2)$, *set*

$$(\mathcal{C}h)_0(z) = (u_z \,|\, \text{Op}(h)u_z),$$
$$(\mathcal{C}h)_1(z) = (u^1_z \,|\, \text{Op}(h)u^1_z) \tag{2.14}$$

for every $z \in \Pi$. *One has the identities* $(\mathcal{C}h)_0 = (\mathcal{C}\,\mathcal{G}h)_0$ *and* $(\mathcal{C}h)_1 = -(\mathcal{C}\,\mathcal{G}h)_1$. *On* Π, *use the standard invariant measure* $d\mu(z) = (\text{Im } z)^{-2}\, d\text{Re } z\, d\text{Im } z$. *If* h *is a* $\mathcal{G}$-*invariant even symbol lying in* $L^2(\mathbb{R}^2)$, *also the image of some function in* $L^2(\mathbb{R}^2)$ *by the operator* $2i\pi\mathcal{E}$, *the function* $(\mathcal{C}h)_0$ *lies in* $L^2(\Pi)$ *and one has*

$$\| (\mathcal{C}h)_0 \|_{L^2(\Pi)} = 2 \,\| \Gamma(i\pi\mathcal{E})\, h \|_{L^2(\mathbb{R}^2)}. \tag{2.15}$$

If h *is an even symbol lying in* $L^2(\mathbb{R}^2)$ *changing to its negative under* $\mathcal{G}$, *the function* $(\mathcal{C}h)_1$ *lies in* $L^2(\Pi)$ *and one has*

$$\| (\mathcal{C}h)_1 \|_{L^2(\Pi)} = 4 \,\| \Gamma(1 + i\pi\mathcal{E})\, h \|_{L^2(\mathbb{R}^2)}. \tag{2.16}$$

Proof. Equation (2.9) makes it clear why even-even (*resp.* odd-odd) symbols have to be used when the corresponding operators are tested against even, or odd, functions only. Equations (2.15) and (2.16) are a special case of Theorem 7.7, and we prefer to substitute for an immediate proof of them an explanation of their significance. In the Weyl calculus, there is coincidence between the L^2-norm of a symbol and the Hilbert-Schmidt norm of the associated operator. This property, more often than not, fails to hold in other quantization procedures, and one of the reasons (it is not the only one) for the introduction, in Chapter 2, of the horocyclic calculus, is to make it true in a more general context.

Consider now the even part of $L^2(\mathbb{R})$, acted upon under the metaplectic representation: one can identify the result with the representation $\pi_{\frac{1}{2}}$ from the discrete series of $\widetilde{G}$, more precisely from the prolongation of this series (*cf.* beginning of Section 6): much more will be said about this when needed. Next, there is another species of symbol one can associate with an endomorphism of the space $H_{\frac{1}{2}}$ of this representation, to wit its Berezin-covariant symbol [3], a certain function on Π. The map $h \mapsto (\mathcal{C}h)_0$ can then be identified with the one which associates with a symbol $h \in L^2_{\text{even}}(\mathbb{R}^2)$ the Berezin-covariant symbol of the operator $\text{Op}(h)$, when regarded as an endomorphism of $H_{\frac{1}{2}}$. One can then see [30, p. 180–181] that, in the case of an even-even symbol,

$$\| h \|^2_{L^2(\mathbb{R}^2)} = (\Lambda^{-1}(\mathcal{C}h)_0 \,|\, (\mathcal{C}h)_0)_{L^2(\Pi)}, \tag{2.17}$$

with

$$\Lambda = 4\Gamma\left(i\sqrt{\Delta - \frac{1}{4}}\right)\Gamma\left(-i\sqrt{\Delta - \frac{1}{4}}\right), \tag{2.18}$$

where $\Delta = -y^2\left(\frac{\partial^2}{\partial x^2} + \frac{\partial^2}{\partial y^2}\right)$ denotes the hyperbolic Laplacian of Π.

One has $z = x+iy = g.i$ if $g = \begin{pmatrix} y^{\frac{1}{2}} & xy^{-\frac{1}{2}} \\ 0 & y^{-\frac{1}{2}} \end{pmatrix}$: then, (2.4) and (2.10), together with the covariance of the Weyl calculus, make it possible to write

$$(\mathcal{C}h)_0(z) = \langle h, W(u_z, u_z) \rangle \tag{2.19}$$

with

$$W(u_z, u_z)(q, p) = W(u_i, u_i)(g^{-1}.(q,p))$$
$$= 2 \exp\left(-\frac{2\pi}{y}|q - zp|^2\right). \tag{2.20}$$

Since this function only depends on $\frac{|q-zp|^2}{y}$, it follows from an immediate calculation [31, p. 17] that, under the map $h \mapsto (\mathcal{C}h)_0$, the operator $\Delta - \frac{1}{4}$ is the transfer of the operator $\pi^2 \mathcal{E}^2$ on $\mathbb{R}^2$, which leads to (2.15). The proof of (2.16) is entirely similar. $\qquad\square$

Remark 2.1. (i) Equations (2.15) and (2.16) are thus just an expression of the way the Berezin symbol map fails to be an isometry.

(ii) One may regard these equations as an intrinsic property of the Berezin symbol map rather than a connection between this map and the Weyl symbol map. Indeed, setting

$$(\mathcal{C}^{\text{op}} A)_0(z) = (u_z \,|\, A\, u_z),$$
$$(\mathcal{C}^{\text{op}} A)_1(z) = (u_z^1 \,|\, A\, u_z^1), \tag{2.21}$$

one can rewrite the equations in a way involving the operator A rather than its Weyl symbol, in the following way: denote as Q the operator acting on functions of $t \in \mathbb{R}$ as the multiplication by t and set $P = \frac{1}{2i\pi}\frac{d}{dt}$. Next, consider the *mixed adjoint* operator $\text{mad}(P \wedge Q)$ acting on Hilbert-Schmidt operators A under the rule

$$\text{mad}(P \wedge Q)(A) = P A Q - Q A P: \tag{2.22}$$

the operator just defined (not a derivation in the algebraic sense) is essentially self-adjoint in the space $H.S.$ of Hilbert-Schmidt operators, if given as an initial domain the space of operators from $\mathcal{S}'(\mathbb{R})$ to $\mathcal{S}(\mathbb{R})$, and one has the spectral-theoretic formula

$$\| (\mathcal{C}^{\mathrm{op}} A)_0 \|_{L^2(\Pi)} = 2 \, \| \, \Gamma \, (i\pi \, \mathrm{mad}(P \wedge Q)) \, A \, \|_{H.S.} \tag{2.23}$$

in the case when A is an even-even operator while, in the case when it is an odd-odd operator, one has

$$\| (\mathcal{C}^{\mathrm{op}} A)_1 \|_{L^2(\Pi)} = 4 \, \| \, \Gamma \, (1 + i\pi \, \mathrm{mad}(P \wedge Q)) \, A \, \|_{H.S.} . \tag{2.24}$$

To obtain this "intrinsic" formulation, it suffices to transfer what has been obtained with the help of the Weyl calculus, using the fact [31, p. 132] that the operators $\mathrm{mad}(P \wedge Q)$ and $\mathcal{E}$ correspond to each other.

There are two essentially different ways to extend the discussion of the maps $h \mapsto (\mathcal{C}h)_0$ and $h \mapsto (\mathcal{C}h)_1$ to an arithmetic environment. The first one consists in making the symbol h an automorphic (say, $SL(2, \mathbb{Z})$-invariant) symbol: this point of view was developed at length in [31] and led to what we called the automorphic pseudodifferential analysis. The second point of view, to be developed in the present chapter, consists in substituting for the families of coherent states (u_z) and (u_z^1) two families (of distributions, as it turns out), in such a way that the parametrizing space should no longer be Π but the quotient of some finite cover of G by an arithmetic group. Besides, everything extends as a one-parameter theory, letting the even and odd parts of the metaplectic representation appear as two special cases of a τ-dependent theory to be introduced in Chapter 2.

3   A family of arithmetic coherent states for the metaplectic representation

We first need to construct distributions $\mathfrak{d}$ on the line invariant, up to multiplication by scalars, under the metaplectic action of some group covering an appropriate subgroup of $\Gamma = SL(2, \mathbb{Z})$. What has to be done is to track the action on $\mathfrak{d}$ of the Fourier transformation, or of the multiplication by $e^{i\pi x^2}$. We first do it from scratch, generalizing the simplest example, to wit the distribution $\mathfrak{d}_0(x) = \sum_{\ell \in \mathbb{Z}} \delta(x - \ell)$, which is invariant under the one-dimensional Fourier transformation $\mathcal{F}$ (2.1) (Poisson's formula). It is also invariant under the multiplication by $e^{2i\pi x^2}$: splitting it into two terms would make it possible to consider the multiplication by $e^{i\pi x^2}$, but would destroy the Fourier invariance.

Theorem 3.1. *Let $\chi^{(12)}$ be the Dirichlet character mod 12 such that $\chi^{(12)}(1) = \chi^{(12)}(11) = 1$, $\chi^{(12)}(5) = \chi^{(12)}(7) = -1$, and let $\chi^{(4)}$ be the Dirichlet character* *mod* 4 *such that* $\chi^{(4)}(1) = 1$, $\chi^{(4)}(3) = -1$. *Set*

$$\mathfrak{d}_{even}(x) = \sum_{m \in \mathbb{Z}} \chi^{(12)}(m)\, \delta\left(x - \frac{m}{\sqrt{12}}\right),$$

$$\mathfrak{d}_{odd}(x) = \sum_{m \in \mathbb{Z}} \chi^{(4)}(m)\, \delta\left(x - \frac{m}{2}\right). \tag{3.1}$$

Any metaplectic transformation above some element of $SL(2, \mathbb{Z})$ acts on the first (resp. second) of these distributions as the multiplication by some 24th (resp. 8th) root of unity.

Proof. Actually, we shall prove more than stated in the theorem: the constructions of the present section depend on some integer N, the value of which will soon be assumed to be 12 or 4. However, in Section 12, we shall only assume that N is 4 times a squarefree integer, an assumption we shall start with here. Given $N \geq 2$, let S be the set of distinct primes dividing N, and let $N = \prod_{p \in S} p^{\alpha_p}$: we assume that $2 \in S$ and $\alpha_2 = 2$, and that $\alpha_p = 1$ for each $p \geq 3$ in S. For every prime $p \in S$, set $N = p^{\alpha_p} M_p$, and let $M = \sum_p M_p$: this number is relatively prime to N, and we choose an integer $\overline{M}$ such that $M\overline{M} \equiv 1$ mod N; in all occurrences of $\overline{M}$, only the (uniquely defined) class of M mod N will be relevant. Set

$$\theta = \exp \frac{2i\pi}{N}, \qquad \theta_p = \exp \frac{2i\pi}{p^{\alpha_p}} = \theta^{M_p}, \tag{3.2}$$

and

$$\iota = \theta^{\overline{M}}, \qquad \iota_p = \theta_p^{\overline{M}} = \theta^{M_p \overline{M}} = \iota^{M_p}. \tag{3.3}$$

One has

$$\theta = \iota^M = \iota^{\sum M_p} = \prod_p \iota_p. \tag{3.4}$$

Consider the multiplicative group $(\mathbb{Z}/N\mathbb{Z})^\times$ of classes mod N relatively prime to N; also, let Λ denote the subgroup $\{\mu \bmod N : \mu^2 \equiv 1 \bmod 2N\}$. Given $j \in \mathbb{Z}$, the class of j mod N will lie in Λ if and only if one has

$$j \equiv \pm 1 \quad \bmod p^{\alpha_p} \quad \text{for every } p. \tag{3.5}$$

Let J be the set consisting of all systems $\kappa = (\kappa_p)_{p \in S}$, with $\kappa_p = \pm 1$ for each $p \in S$. The map $\mu \mapsto \kappa$ from Λ to J characterized by the validity of the congruence $\mu \equiv \kappa_p \bmod p^{\alpha_p}$ for each p is one-to-one: it is an isomorphism once J has been identified with a product of groups (one for each $p \in S$), each of which coincides with $\mathbb{Z}/2\mathbb{Z}$.

Denoting as ϕ_p the non-trivial character of the group corresponding to p in the decomposition of J as a product of groups $\mathbb{Z}/2\mathbb{Z}$, set

$$\phi(\kappa) = \prod_{p \in S} \phi_p(\kappa_p). \tag{3.6}$$

and let χ be the character: $\Lambda \to \{\pm 1\}$ corresponding to ϕ under the natural isomorphism from J to Λ. If $\mu \in \Lambda$ is associated to $\kappa \in J$, one has for every $j \in \mathbb{Z}$, using (3.4) (and noting that θ^μ makes sense for $\mu \in \mathbb{Z}$ only defined mod N),

$$\sum_{\mu \in \Lambda} \chi(\mu) \, \theta^{j\mu} = \sum_{\kappa \in J} \phi(\kappa) \prod_p \iota_p^{j\kappa_p}$$

$$= \sum_{\kappa \in J} \prod_{p \in S} \left(\phi_p(\kappa_p) \, \iota_p^{j\kappa_p} \right)$$

$$= \prod_{p \in S} \left(\sum_{\kappa_p = \pm 1} \phi_p(\kappa_p) \, \iota_p^{j\kappa_p} \right)$$

$$= \prod_{p \in S} \left(\iota_p^{j} - \iota_p^{-j} \right) . \tag{3.7}$$

Now, assume that j has a common prime factor p with N: then, under the assumptions relative to N, p^{α_p} divides $2j$, so that the corresponding factor $\iota_p^{j} - \iota_p^{-j}$ of the product just written is zero. Consequently, one has

$$\sum_{\mu \in \Lambda} \chi(\mu) \, \theta^{\mu\nu} = 0 \tag{3.8}$$

for every $\nu \in \mathbb{Z}/N\mathbb{Z}$ not in $(\mathbb{Z}/N\mathbb{Z})^\times$.

Lemma 3.2. *With the notation just made clear, assume that $\alpha_2 = 2$ and that $\alpha_p = 1$ for every $p \geq 3$ in S, and let R_N be a fixed set of representatives of $(\mathbb{Z}/N\mathbb{Z})^\times$ mod Λ. For each $\rho \in R_N$, consider the distribution*

$$\varpi_\rho(x) = \sum_{\mu \in \Lambda} \chi(\mu) \sum_{\ell \in \mathbb{Z}} \delta \left(x - \frac{N\ell + \rho\mu}{\sqrt{N}} \right) . \tag{3.9}$$

As ρ describes R_N, these distributions are linearly independent. The space V they generate is left invariant under the multiplication by $e^{i\pi x^2}$ as well as under the inverse Fourier transformation. The matrices representing these two transformations in the basis so defined are respectively the diagonal matrix T with entries $e^{\frac{i\pi\rho^2}{N}}$ and the matrix K with entries

$$K(\rho, \sigma) = N^{-\frac{1}{2}} \prod_{p \in S} \left(\iota_p^{\rho\sigma} - \iota_p^{-\rho\sigma} \right) . \tag{3.10}$$

Proof. First note that ϖ_ρ depends on the choice of the representative ρ in the group $(\mathbb{Z}/N\mathbb{Z})^\times / \Lambda$ in the following way: if $\lambda \in \Lambda$, one has $\varpi_{\lambda\rho} = \chi(\lambda) \, \varpi_\rho$. It is immediate (it is at this point that the assumption that N is even, which eventually led to $\alpha_2 = 2$, is needed) that

$$e^{i\pi x^2} \varpi_\rho(x) = e^{\frac{i\pi\rho^2}{N}} \varpi_\rho(x). \tag{3.11}$$

Next, we consider the effect on ϖ_ρ of the inverse Fourier transformation. Since, in view of Poisson's formula,

$$\mathcal{F}^{-1} \sum_{\ell \in \mathbb{Z}} \delta \left(x - \frac{N\ell + \rho\mu}{\sqrt{N}} \right) = N^{-\frac{1}{2}} \sum_{m \in \mathbb{Z}} e^{\frac{2i\pi m\rho\mu}{N}} \delta \left(x - \frac{m}{\sqrt{N}} \right), \tag{3.12}$$

one has

$$(\mathcal{F}^{-1} \varpi_\rho)(x) = N^{-\frac{1}{2}} \sum_{\mu \in \Lambda} \chi(\mu) \sum_{m \in \mathbb{Z}} \theta^{m\rho\mu} \delta \left(x - \frac{m}{\sqrt{N}} \right). \tag{3.13}$$

Setting $m = N\ell + \nu$, $\nu \bmod N$, $\ell \in \mathbb{Z}$, one finds

$$(\mathcal{F}^{-1} \varpi_\rho)(x) = N^{-\frac{1}{2}} \sum_{\mu \in \Lambda} \chi(\mu) \sum_{\nu \bmod N} \sum_{\ell \in \mathbb{Z}} \theta^{\rho\mu\nu} \delta \left(x - \frac{N\ell + \nu}{\sqrt{N}} \right). \tag{3.14}$$

Now, (3.8) shows that $\sum_{\mu \in \Lambda} \chi(\mu)\, \eta^{\rho\mu\nu} = 0$ unless $\nu \in (\mathbb{Z}/N\mathbb{Z})^\times$, so that we may restrict the sum relative to ν to this latter range: then, we replace ν by $\sigma\nu$, with $(\sigma, \nu) \in R_N \times \Lambda$, getting

$$(\mathcal{F}^{-1} \varpi_\rho)(x) = N^{-\frac{1}{2}} \sum_{\sigma \in R_N} \sum_{\mu \in \Lambda} \sum_{\nu \in \Lambda} \sum_{\ell \in \mathbb{Z}} \chi(\mu)\, \theta^{\rho\sigma\mu\nu} \delta \left(x - \frac{N\ell + \sigma\nu}{\sqrt{N}} \right). \tag{3.15}$$

Since $\nu^{-1} = \nu$ for $\nu \in \Lambda$, one has, using (3.7),

$$\sum_{\mu \in \Lambda} \chi(\mu)\, \theta^{\rho\sigma\mu\nu} = \sum_{\mu \in \Lambda} \chi(\mu\nu)\, \theta^{\rho\sigma\mu}$$

$$= \chi(\nu) \sum_{\mu \in \Lambda} \chi(\mu)\, \theta^{\rho\sigma\mu}$$

$$= \chi(\nu) \prod_{p \in S} \left(\iota_p^{\rho\sigma} - \iota_p^{-\rho\sigma} \right). \tag{3.16}$$

Hence,

$$(\mathcal{F}^{-1} \varpi_\rho)(x) = N^{-\frac{1}{2}} \sum_{\sigma \in R_N} \sum_{\nu \in \Lambda} \sum_{\ell \in \mathbb{Z}} \chi(\nu) \prod_{p \in S} \left(\iota_p^{\rho\sigma} - \iota_p^{-\rho\sigma} \right) \delta \left(x - \frac{N\ell + \sigma\nu}{\sqrt{N}} \right). \tag{3.17}$$

In other words, defining $K(\rho, \sigma)$ by (3.10), one has

$$\mathcal{F}^{-1} \varpi_\rho = \sum_{\sigma \in R_N} K(\rho, \sigma)\, \varpi_\sigma. \tag{3.18}$$

$\square$

End of proof of Theorem 3.1. Note that the distribution ϖ_ρ is even (*resp.* odd) according to the parity of the number $\#S$ of elements of S. On the other hand, the entries of the symmetric matrix K are real or pure imaginary according to whether $\#S$ is even or odd: since $(\mathcal{F}^2\varpi)(x) = \varpi(-x)$ for every distribution ϖ, it follows that K is an orthogonal symmetric matrix or i times such a matrix according to the parity of $\#S$. This provides two orthogonal subspaces of V, on which $\mathcal{F}^{-1}$ acts as the pair of scalars $(1, -1)$ or $(i, -i)$. Finding within V a basis of eigenvectors of the Fourier transformation is then possible. However, the matrices K and T do not have a single common eigenvector unless T is a scalar: with the aim of finding a tempered distribution invariant, up to some phase factor, under every metaplectic transformation lying above some element of $SL(2, \mathbb{Z})$, we are left only with the situation when $\#R_N = 1$, i.e., when $\Lambda = (\mathbb{Z}/N\mathbb{Z})^\times$. Hence, only the cases when $N = 12$ or 4 need be examined.

One has $M = 7 = \overline{M}$ in the first case, $M = 1$ in the second one. The group J is isomorphic to $\mathbb{Z}/2\mathbb{Z} \times \mathbb{Z}/2\mathbb{Z}$ or $\mathbb{Z}/2\mathbb{Z}$. The character χ of Λ is given by the following formulas: when $N = 12$, $\chi(1) = \chi(11) = 1$, $\chi(5) = \chi(7) = -1$; when $N = 4$, $\chi(1) = 1$, $\chi(3) = -1$. Indeed, in the first case, the elements $1, 5, 7, 11$ of $(\mathbb{Z}/12\mathbb{Z})^\times$ correspond, under the isomorphism above, to the pairs $(1, 1), (1, 2), (3, 1), (3, 2)$ in the group $(\mathbb{Z}/4\mathbb{Z})^\times \times (\mathbb{Z}/3\mathbb{Z})^\times$. The set R_N reduces to one element, of course chosen as the unit element of $(\mathbb{Z}/N\mathbb{Z})^\times$, so we abbreviate ϖ_ρ as $\mathfrak{d}^{(N)}$, keeping track of N instead: it is given in the two cases as indicated in (3.1).

The number T (formerly a matrix) is $e^{\frac{i\pi}{12}}$ or $e^{\frac{i\pi}{4}}$ in the two cases. The number K (formerly a matrix too) has the respective values

$$12^{-\frac{1}{2}} \left(e^{\frac{7i\pi}{2}} - e^{-\frac{7i\pi}{2}}\right)\left(e^{\frac{14\,i\pi}{3}} - e^{-\frac{14\,i\pi}{3}}\right) = 1, \tag{3.19}$$

$$\frac{1}{2}\left(e^{\frac{i\pi}{2}'} - e^{-\frac{i\pi}{2}}\right) = i. \qquad \square$$

The Dirichlet L-functions associated with the characters $\chi^{(12)}$ and $\chi^{(4)}$ act as spectral densities, when $\mathfrak{d}^{(12)} = \mathfrak{d}_{\text{even}}$ and $\mathfrak{d}^{(4)} = \mathfrak{d}_{\text{odd}}$ are decomposed into their homogeneous components.

Proposition 3.3. *With* $L(s, \chi^{(N)}) = \sum_{m\geq 1} \frac{\chi^{(N)}(m)}{m^s}$ *for* Re $s > 1$, *and denoting in the same way the holomorphic extension of this function to the complex plane, one has*

$$\mathfrak{d}_{\text{even}}(x) = \int_{-\infty}^{\infty} c_{\text{even}}(\lambda)\, |x|^{-\frac{1}{2}-i\lambda}\, d\lambda,$$

$$\mathfrak{d}_{\text{odd}}(x) = \int_{-\infty}^{\infty} c_{\text{odd}}(\lambda)\, |x|^{-\frac{1}{2}-i\lambda}\, \text{sign}\, x\, d\lambda \tag{3.20}$$

with

$$c_{\text{even}}(\lambda) = \frac{1}{2\pi}\, (12)^{\frac{1}{4}-\frac{i\lambda}{2}}\, L\left(\frac{1}{2} - i\lambda, \chi^{(12)}\right),$$

$$c_{\text{odd}}(\lambda) = \frac{1}{2\pi}\, 2^{\frac{1}{2}-i\lambda}\, L\left(\frac{1}{2} - i\lambda, \chi^{(4)}\right). \tag{3.21}$$

Proof. First note that (3.1) can also be written

$$\mathfrak{d}_{\text{even}}(x) = \sum_{m \geq 1} \chi^{(12)}(m) \left[\delta\left(x - \frac{m}{\sqrt{12}} \right) + \delta\left(x + \frac{m}{\sqrt{12}} \right) \right],$$

$$\mathfrak{d}_{\text{odd}}(x) = \sum_{m \geq 1} \chi^{(4)}(m) \left[\delta\left(x - \frac{m}{2} \right) - \delta\left(x + \frac{m}{2} \right) \right]. \tag{3.22}$$

Let us prove the first decomposition (3.20) only, since the second one is obtained in an entirely similar way. If $u \in \mathcal{S}_{\text{even}}(\mathbb{R})$, the Schwartz space of even, smooth functions on the line, rapidly decreasing at infinity, and if one sets

$$\phi(s) = \frac{1}{2\pi} \int_0^\infty t^{s-1} u(t) \, dt, \quad \text{Re } s > 0, \tag{3.23}$$

obtaining a function holomorphic in this half-plane, rapidly decreasing at infinity on vertical lines, one has

$$u(x) = \frac{1}{i} \int_{\sigma - i\infty}^{\sigma + i\infty} |x|^{-s} \phi(s) \, ds \tag{3.24}$$

for $\sigma > 0$, an identity which becomes the spectral decomposition of u relative to the self-adjoint operator $i\left(\frac{1}{2} + x \frac{d}{dx} \right)$ in the case when $\sigma = \frac{1}{2}$.

For $\sigma > 1$, one can then write

$$
\begin{aligned}
\langle \mathfrak{d}_{\text{even}}, u \rangle &= \sum_{m \geq 1} \chi^{(12)}(m) \, \frac{2}{i} \int_{\sigma - i\infty}^{\sigma + i\infty} \left(\frac{m}{\sqrt{12}} \right)^{-s} \phi(s) \, ds \\
&= \frac{2}{i} \int_{\sigma - i\infty}^{\sigma + i\infty} (12)^{\frac{s}{2}} L(s, \chi^{(12)}) \, \phi(s) \, ds :
\end{aligned}
\tag{3.25}
$$

in view of the well-known extension of Dirichlet L-functions as entire functions, one can put back σ to the value $\frac{1}{2}$; comparing this identity to the equation

$$\langle \mathfrak{d}_{\text{even}}, u \rangle = 4\pi \int_{-\infty}^\infty c_{\text{even}}(\lambda) \phi\left(\frac{1}{2} - i\lambda \right) d\lambda, \tag{3.26}$$

just another way to write (3.20), we obtain the expression of $c_{\text{even}}(\lambda)$.   □

Remark 3.1. Since

$$\mathcal{F}\left(|x|^{-\frac{1}{2} - i\lambda} \right) = \pi^{i\lambda} \frac{\Gamma\left(\frac{1}{4} - \frac{i\lambda}{2} \right)}{\Gamma\left(\frac{1}{4} - \frac{i\lambda}{2} \right)} |x|^{-\frac{1}{2} + i\lambda},$$

$$\mathcal{F}\left(|x|^{-\frac{1}{2} - i\lambda} \operatorname{sign} x \right) = -i \, \pi^{i\lambda} \frac{\Gamma\left(\frac{3}{4} - \frac{i\lambda}{2} \right)}{\Gamma\left(\frac{3}{4} - \frac{i\lambda}{2} \right)} |x|^{-\frac{1}{2} + i\lambda} \operatorname{sign} x, \tag{3.27}$$

the equations

$$\mathcal{F}^{-1} \mathfrak{d}_{\text{even}} = \mathfrak{d}_{\text{even}}, \qquad \mathcal{F}^{-1} \mathfrak{d}_{\text{odd}} = i\,\mathfrak{d}_{\text{odd}} \tag{3.28}$$

yield

$$\pi^{\frac{i\lambda}{2}} \Gamma\left(\frac{1}{4} - \frac{i\lambda}{2}\right) c_{\text{even}}(\lambda) = \pi^{-\frac{i\lambda}{2}} \Gamma\left(\frac{1}{4} + \frac{i\lambda}{2}\right) c_{\text{even}}(-\lambda),$$

$$\pi^{\frac{i\lambda}{2}} \Gamma\left(\frac{3}{4} - \frac{i\lambda}{2}\right) c_{\text{odd}}(\lambda) = \pi^{-\frac{i\lambda}{2}} \Gamma\left(\frac{3}{4} + \frac{i\lambda}{2}\right) c_{\text{odd}}(-\lambda). \tag{3.29}$$

In view of (3.21), the first equation is equivalent to the invariance, under the transformation $s \mapsto 1-s$, of the function $\left(\frac{12}{\pi}\right)^{\frac{s}{2}} \Gamma(\frac{s}{2}) L(s, \chi^{(12)})$, and the second one is equivalent to the invariance under the same transformation of the function $\left(\frac{4}{\pi}\right)^{\frac{s}{2}} \Gamma(\frac{1+s}{2}) L(s, \chi^{(4)})$. In other words [6, p. 152] or [12, p. 204], once Proposition 3.3 has been established, the way $\mathfrak{d}_{\text{even}}$, or $\mathfrak{d}_{\text{odd}}$, transforms under the Fourier transformation, is equivalent to the functional equation of the associated Dirichlet L-function.

The game can be played in reverse, as has been shown in [32, p. 1165]. Starting with a modular form of an appropriate species, and using the functional equation of the associated Dirichlet series, one can build miscellaneous formulas, generalizing Poisson's or Voronoi's. These generalizations are concerned with d-dimensional combs (*i.e.*, $SL(d, \mathbb{Z})$-invariant measures supported in $\mathbb{Z}^d$), where one is free to choose d; on the other hand, it is not exactly the Fourier transformation that occurs in these formulas, rather the product of this transformation by some function, in the spectral-theoretic sense, of the d-dimensional Euler operator.

We have done our best to present the construction of $\mathfrak{d}_{\text{even}}$ and $\mathfrak{d}_{\text{odd}}$ while staying entirely within the usual realization of the metaplectic representation. It is, however, necessary to connect the result to the theory of modular forms of weights $\frac{1}{2}$ and $\frac{3}{2}$. To this effect, we need to introduce in succession two intertwining operators: the first one is a quadratic transformation Sq_{even} or Sq_{odd} of which (in the present work) we shall not have to consider any more general version. The second one is a Laplace transformation, from some weighted L^2-space of functions on $(0, \infty)$ to some space of holomorphic functions in Π. As such, it is a special case of a one-parameter family which will have to be introduced in Section 6. Anticipating the exposition in that section, we consider the spaces $H_{\frac{1}{2}} = L^2((0, \infty), t^{\frac{1}{2}} dt)$ and $H_{\frac{3}{2}} = L^2((0, \infty), t^{-\frac{1}{2}} dt)$: it is immediate that the map $\text{Sq}_{\text{even}} \colon v \mapsto u$, with $u(x) = 2^{-\frac{3}{4}} |x|\, v(\frac{x^2}{2})$, is an isometry from $H_{\frac{1}{2}}$ onto $L^2_{\text{even}}(\mathbb{R})$, and that the map $\text{Sq}_{\text{odd}} \colon v \mapsto u$, with $u(x) = 2^{-\frac{1}{4}} v(\frac{x^2}{2}) \operatorname{sign} x$, is an isometry from $H_{\frac{3}{2}}$ onto $L^2_{\text{odd}}(\mathbb{R})$. Note that these two quadratic transformations continue to make sense on the distribution level, provided 0 does not lie in the support of the distribution under consideration: if $a > 0$, one has $\delta(x-a) = v(\frac{x^2}{2})$ if $v(t) = a\,\delta(t - \frac{a^2}{2})$.

It will be recalled in the beginning of Section 6 that the map $\mathrm{Sq}_{\mathrm{even}}$ (*resp.* $\mathrm{Sq}_{\mathrm{odd}}$) intertwines the even (*resp.* odd) part of the metaplectic representation with a representation denoted as $\mathcal{D}_{\frac{1}{2}}$ (*resp.* $\mathcal{D}_{\frac{3}{2}}$) in the Hilbert space $H_{\frac{1}{2}}$ (*resp.* $H_{\frac{3}{2}}$). It is to be noted that the representations $\mathcal{D}_{\frac{1}{2}}$ and $\mathcal{D}_{\frac{3}{2}}$, just like the metaplectic representation, are genuine representations of the twofold cover $\widetilde{G}$ of $G = SL(2,\mathbb{R})$, not of G: however, if one agrees to let the indeterminacy factor ± 1 subsist in the formulas, one may consider these as *projective* representations of $SL(2,\mathbb{R})$. Let us take this opportunity to mention that – though we shall sometimes have to worry about phase factors – projective representations, *i.e.*, representations up to indeterminate phase factors, will be just as good for our purposes as genuine representations, since operators from such representations will most often occur on both sides of scalar products. There is only one advantage in considering a representation of $\widetilde{G}$ as a representation, up to an indeterminate factor ± 1, of G: it is the fact that it always demands some effort to simply *name* an element of $\widetilde{G}$.

Forgetting the constant $2^{-\frac{3}{4}}$ or $2^{-\frac{1}{4}}$, which only served a normalization purpose, we may transform the distribution $\mathfrak{d}^{(N)}$, in the two cases under discussion, into the distribution $\mathfrak{S}^{(N)}$ on the half-line, considered as lying in a space extending $H_{\frac{1}{2}}$ in the first case, extending $H_{\frac{3}{2}}$ in the second one, given as follows:

$$\mathfrak{S}^{(12)}(t) = \sum_{m \geq 1} \chi^{(12)}(m)\, \delta\left(t - \frac{m^2}{24}\right), \tag{3.30}$$

or

$$\mathfrak{S}^{(4)}(t) = \sum_{m \geq 1} \chi^{(4)}(m)\, \delta\left(t - \frac{m^2}{8}\right). \tag{3.31}$$

When all representations, in the current context, are considered as projective representations of G, the metaplectic transformations associated to the matrices $\left(\begin{smallmatrix} 0 & 1 \\ -1 & 0 \end{smallmatrix}\right)$ and $\left(\begin{smallmatrix} 1 & 0 \\ 1 & 1 \end{smallmatrix}\right)$ are respectively $\pm e^{-\frac{i\pi}{4}}\,\mathcal{F}$ and $\pm (e^{i\pi x^2})$. Hence, the formulas given above regarding the scalars K and T can be written

$$\mathrm{Met}\left(\left(\begin{smallmatrix} 0 & 1 \\ -1 & 0 \end{smallmatrix}\right)\right) \mathfrak{d}^{(12)} = \pm e^{\frac{i\pi}{4}}\,\mathfrak{d}^{(12)}, \qquad \mathrm{Met}\left(\left(\begin{smallmatrix} 1 & 0 \\ 1 & 1 \end{smallmatrix}\right)\right) \mathfrak{d}^{(12)} = \pm e^{\frac{i\pi}{12}}\,\mathfrak{d}^{(12)},$$

$$\mathrm{Met}\left(\left(\begin{smallmatrix} 0 & 1 \\ -1 & 0 \end{smallmatrix}\right)\right) \mathfrak{d}^{(4)} = \pm e^{-\frac{i\pi}{4}}\,\mathfrak{d}^{(4)}, \qquad \mathrm{Met}\left(\left(\begin{smallmatrix} 1 & 0 \\ 1 & 1 \end{smallmatrix}\right)\right) \mathfrak{d}^{(4)} = \pm e^{\frac{i\pi}{4}}\,\mathfrak{d}^{(4)}. \tag{3.32}$$

These formulas transfer to

$$\mathcal{D}_{\frac{1}{2}}\left(\left(\begin{smallmatrix} 0 & 1 \\ -1 & 0 \end{smallmatrix}\right)\right) \mathfrak{S}^{(12)} = \pm e^{\frac{i\pi}{4}}\,\mathfrak{S}^{(12)}, \qquad \mathcal{D}_{\frac{1}{2}}\left(\left(\begin{smallmatrix} 1 & 0 \\ 1 & 1 \end{smallmatrix}\right)\right) \mathfrak{S}^{(12)} = \pm e^{\frac{i\pi}{12}}\,\mathfrak{S}^{(12)},$$

$$\mathcal{D}_{\frac{3}{2}}\left(\left(\begin{smallmatrix} 0 & 1 \\ -1 & 0 \end{smallmatrix}\right)\right) \mathfrak{S}^{(4)} = \pm e^{-\frac{i\pi}{4}}\,\mathfrak{S}^{(4)}, \qquad \mathcal{D}_{\frac{3}{2}}\left(\left(\begin{smallmatrix} 1 & 0 \\ 1 & 1 \end{smallmatrix}\right)\right) \mathfrak{S}^{(4)} = \pm e^{\frac{i\pi}{4}}\,\mathfrak{S}^{(4)}, \tag{3.33}$$

it being understood that, in the latter formulas, we make use of extensions of the representations involved (a priori, unitary representations) to some distribution setting.

Next, we use another intertwining operator, towards the space of the representation $\pi_{\frac{1}{2}}$ or $\pi_{\frac{3}{2}}$ from the discrete series of the twofold cover of $SL(2,\mathbb{R})$ (the metaplectic group): to be precise, the first of these belongs properly to the prolongation of the discrete series of, say, the universal cover of G (cf. beginning of Section 6). Such an intertwining operator $v \mapsto f$ is given by the formula [31, p. 60]

$$f(z) = \pm e^{\frac{i\pi}{4}} z^{-\frac{1}{2}} \int_0^\infty v(t) e^{-2i\pi t z^{-1}} dt \qquad (3.34)$$

in the first case, and

$$f(z) = \pm e^{-\frac{i\pi}{4}} z^{-\frac{3}{2}} \int_0^\infty v(t) e^{-2i\pi t z^{-1}} dt \qquad (3.35)$$

in the second one. The coefficients, to be decided in full later, have been chosen for convenience, and differ from the one given (in the second case only, for a normalization purpose) in the above reference: this, of course, does not destroy the intertwining property. Again, the representations $\pi_{\frac{1}{2}}$ and $\pi_{\frac{3}{2}}$ are genuine representations of $\widetilde{G}$, but the indeterminacy factors ± 1 are of no concern to us here, so we may recall the definitions of these two representations, viewed as projective representations of $SL(2,\mathbb{R})$, as follows: given $g = \left(\begin{smallmatrix} a & b \\ c & d \end{smallmatrix} \right)$,

$$\left(\pi_{\frac{1}{2}}(g) f \right)(z) = \pm (-cz + a)^{-\frac{1}{2}} f\left(\frac{dz - b}{-cz + a} \right),$$

$$\left(\pi_{\frac{3}{2}}(g) f \right)(z) = \pm (-cz + a)^{-\frac{3}{2}} f\left(\frac{dz - b}{-cz + a} \right). \qquad (3.36)$$

Let $f^{(12)}$ and $f^{(4)}$ be the transforms of $\mathfrak{S}^{(12)}$ and $\mathfrak{S}^{(4)}$ under the transformations $(3.34), (3.35)$ and (3.35) respectively. In view of the left-hand column of the set of equations (3.33), one has

$$\pi_{\frac{1}{2}} \left(\left(\begin{smallmatrix} 0 & 1 \\ -1 & 0 \end{smallmatrix} \right) \right) f^{(12)} = \pm e^{\frac{i\pi}{4}} f^{(12)},$$

$$\pi_{\frac{3}{2}} \left(\left(\begin{smallmatrix} 0 & 1 \\ -1 & 0 \end{smallmatrix} \right) \right) f^{(4)} = \pm e^{-\frac{i\pi}{4}} f^{(4)}. \qquad (3.37)$$

Using (3.36) and (3.37), one obtains the formulas, slightly simpler than (3.34) and (3.35), and correct for appropriate choices of the signs $\pm$ there:

$$f^{(12)}(z) = \int_0^\infty \mathfrak{S}^{(12)}(t) e^{2i\pi t z} dt, \qquad f^{(4)}(z) = \int_0^\infty \mathfrak{S}^{(4)}(t) e^{2i\pi t z} dt. \qquad (3.38)$$

Note that it would have been impossible to define the map $v \mapsto f$ by such a simpler expression in general, in view of the exact desired intertwining property: pseudodifferential analysis (including the version of the metaplectic representation suitable in this context) and representation theory have been developed as independent trades. If one combines the map $u \mapsto v$ from $L^2_{\text{even}}(\mathbb{R})$ to $H_{\frac{1}{2}}$ (resp. from

$L^2_{\text{odd}}(\mathbb{R})$ to $H_{\frac{3}{2}}$) characterized by $u(x) = |x| \, v(\frac{x^2}{2})$ (*resp.* $u(x) = v(\frac{x^2}{2}) \operatorname{sign} x$) with the Laplace transformation that occurs in (3.38), one obtains

$$
\begin{cases}
f(z) = \frac{1}{2} \int_{-\infty}^{\infty} u(x) \, e^{i\pi z x^2} \, dx & \text{in the even case}, \\
f(z) = \frac{1}{2} \int_{-\infty}^{\infty} x \, u(x) \, e^{i\pi z x^2} \, dx & \text{in the odd case}.
\end{cases}
\tag{3.39}
$$

Consequently, if one starts, as in the present case, with discrete sums of Dirac masses on the line, one will find for f a theta-series.

Theorem 3.4. *Transferring* $\mathfrak{d}_{\text{even}} = \mathfrak{d}^{(12)}$ *(resp.* $\mathfrak{d}_{\text{odd}} = \mathfrak{d}^{(4)}$*) to* $\mathfrak{S}^{(12)}$ *(resp.* $\mathfrak{S}^{(4)}$*) under the quadratic transform* $2^{\frac{3}{4}} \operatorname{Sq}_{\text{even}}$ *(resp.* $2^{\frac{1}{4}} \operatorname{Sq}_{\text{odd}}$*), next to* $f^{(12)}$ *(resp.* $f^{(4)}$*) through the two versions of the Laplace transformation in (3.38), one obtains*

$$
f^{(12)} = \eta, \qquad f^{(4)} = \frac{1}{2} \eta^3,
\tag{3.40}
$$

where η *is Dedekind's eta function,* $(2\pi)^{-\frac{1}{2}}$ *times a 24th root of Ramanujan's Delta function.*

Proof. Setting, as is classically done, $q = e^{2i\pi z}$, we obtain for $f^{(12)}$ the expansion, in which m describes the part of $\mathbb{Z}$ subject to the restriction indicated:

$$
f^{(12)}(z) = \frac{1}{2} \left[\sum_{m \equiv 1 \bmod 12} q^{\frac{m^2}{24}} + \sum_{m \equiv 11 \bmod 12} q^{\frac{m^2}{24}} - \sum_{m \equiv 5 \bmod 12} q^{\frac{m^2}{24}} - \sum_{m \equiv 7 \bmod 12}{}' q^{\frac{m^2}{24}} \right]
$$

$$
= \frac{1}{2} \sum_{\varepsilon = \pm 1} \left[\sum_{\ell \in \mathbb{Z}} q^{\frac{(12\ell+\varepsilon)^2}{24}} - \sum_{\ell \in \mathbb{Z}} q^{\frac{(12\ell+6+\varepsilon)^2}{24}} \right]
$$

$$
= \frac{1}{2} \sum_{\varepsilon = \pm 1} \sum_{\ell \in \mathbb{Z}} (-1)^{\ell} \, q^{\frac{(6\ell-\varepsilon)^2}{24}}
$$

$$
= \frac{1}{2} q^{\frac{1}{24}} \sum_{\varepsilon = \pm 1} \sum_{\ell \in \mathbb{Z}} (-1)^{\ell} \, q^{\frac{\ell(3\ell-\varepsilon)}{2}}
\tag{3.41}
$$

or, finally,

$$
f^{(12)}(z) = q^{\frac{1}{24}} \sum_{\ell \in \mathbb{Z}} (-1)^{\ell} \, q^{\frac{\ell(3\ell-1)}{2}}.
\tag{3.42}
$$

One recognizes Euler's pentagonal coefficients as they occur in the identity

$$
\prod_{n \geq 1} (1 - q^n) = \sum_{\ell \in \mathbb{Z}} (-1)^{\ell} \, q^{\frac{\ell(3\ell-1)}{2}}
\tag{3.43}
$$

and, as shown by this product expansion [19, p. 134], $f^{(12)}$ coincides with the Dedekind eta function. In particular, the 24th power of this function coincides with $(2\pi)^{-12}$ times Ramanujan's Δ function.

In the same way,

$$f^{(4)}(z) = \sum_{\ell \geq 0} \left[\frac{4\ell + 1}{2} q^{\frac{(4\ell+1)^2}{8}} - \frac{4\ell + 3}{2} q^{\frac{(4\ell+3)^2}{8}} \right]$$

$$= q^{\frac{1}{8}} \sum_{\ell \geq 0} \left[\frac{4\ell + 1}{2} q^{\ell(2\ell+1)} - \frac{4\ell + 3}{2} q^{(\ell+1)(2\ell+1)} \right]$$

$$= \frac{1}{2} q^{\frac{1}{8}} \sum_{n \geq 0} (-1)^n (2n + 1) q^{\frac{n(n+1)}{2}} . \qquad (3.44)$$

This is a modular form of sorts of weight $\frac{3}{2}$ for the full modular group: only, indeterminacy coefficients, all lying in the group generated by $e^{\frac{i\pi}{4}}$, occur in the transformation formulas; also, powers of $q^{\frac{1}{8}}$ occur in the Fourier expansion. Taking the 8th power solves all problems at once. The function $(f^{(4)})^8$ is a genuine cusp-form for the full modular group of weight $\frac{3}{2} \times 8 = 12$: again, it is a multiple of the Ramanujan Δ function, of necessity (taking the constant term of the Fourier expansion into consideration) 2^{-8} times $(f^{(12)})^{24}$. This gives the identity

$$\sum_{n \geq 0} (-1)^n (2n + 1) q^{\frac{n(n+1)}{2}} = \left[\sum_{\ell \in \mathbb{Z}} (-1)^\ell q^{\frac{\ell(3\ell-1)}{2}} \right]^3$$

$$= \prod_{n \geq 1} (1 - q^n)^3 , \qquad (3.45)$$

this time a special case of an identity of Jacobi [19, p. 172], originating from the theory of θ-functions and usually put to use in the proof of Ramanujan's identities for the partition function. Its significance for our purpose is that it is associated with an odd distribution $\mathfrak{d}^{(4)}$ on the line, supported in the set $\frac{1}{2} + \mathbb{Z}$, which is an eigendistribution of the Fourier transformation as well as of the multiplication by $e^{i\pi x^2}$; the even distribution $\mathfrak{d}^{(12)}$, supported in the set $\{ \frac{m}{\sqrt{12}} : m \equiv 1, 5, 7, 11 \bmod 12 \}$, has the same property. $\square$

If one agrees to replace the full modular group by some subgroup, there are more possibilities, from the most obvious distribution $\mathfrak{d}_0(x) = \sum_{\ell \in \mathbb{Z}} \delta(x - \ell)$, which satisfies the same kind of properties with respect to the group generated by $\begin{pmatrix} 0 & 1 \\ -1 & 0 \end{pmatrix}$ and $\begin{pmatrix} 1 & 0 \\ 2 & 1 \end{pmatrix}$, a group isomorphic to Hecke's subgroup $\Gamma_0(2)$. Working things in reverse, i.e., starting from the modular form f, one might find other possibilities, but we shall concentrate, in the remainder of this chapter, on the distributions which can be obtained from the consideration of the distributions ϖ_ρ.

4   Wigner functions of pairs of arithmetic coherent states

We compute here the Wigner function, in the sense of the Weyl calculus, of a diagonal pair $(\mathfrak{d}, \mathfrak{d})$ in which $\mathfrak{d}$ is one of the arithmetic distributions $\mathfrak{d}_{even}, \mathfrak{d}_{odd}, \mathfrak{d}_0$, singled out in Section 3. Something can be done too with the distributions ϖ_ρ for general N (cf. Lemma 3.2) provided that one consider all ϖ_ρ's simultaneously: we shall only come to the general situation in Section 12.

Also, we try to understand more about the transforms of each of the first two distributions under an arbitrary metaplectic transformation. It would be very hard to analyze exactly the effect on a distribution such as ϖ_ρ of an arbitrary metaplectic transformation lying above some element of $SL(2, \mathbb{Z})$. Even in the case when $N = 12$, so that $\varpi = \mathfrak{d}_{even}$, finding the exact phase factor is a problem tantamount to that of finding the exact transformation rule satisfied by the Dedekind η-function: though its solution goes back to the 19th century, it is by no means easy [19, p. 163] and involves Legendre-Jacobi symbols. Moving to Wigner functions suppresses many difficulties: of course, it also lessens the information since, in the case of a diagonal pair, it is indifferent to phase factors.

Set $\mathfrak{d}_{even}^{\tilde{g}} = \mathrm{Met}(\tilde{g}^{-1})\,\mathfrak{d}_{even}$ and $\mathfrak{d}_{odd}^{\tilde{g}} = \mathrm{Met}(\tilde{g}^{-1})\,\mathfrak{d}_{odd}$ for every $\tilde{g} \in \widetilde{G}$: recall that the distribution obtained only depends, up to some phase factor, on the point $g \in G$ above which $\tilde{g}$ lies. Hence, given $u \in \mathcal{S}(\mathbb{R})$, the absolute value $|\langle \mathfrak{d}_{even}^{\tilde{g}}, u \rangle|$ can be denoted as $|\langle \mathfrak{d}_{even}^{g}, u \rangle|$, and the same goes with odd distributions. Normalizing the Haar measure on G by the equation $dg = \frac{1}{2\pi}\,d\mu(z)\,d\theta$ if $z = x + iy$ and $g = \begin{pmatrix} y^{\frac{1}{2}} & xy^{-\frac{1}{2}} \\ 0 & y^{-\frac{1}{2}} \end{pmatrix} \begin{pmatrix} \cos\theta & -\sin\theta \\ \sin\theta & \cos\theta \end{pmatrix}$, $0 \le \theta < 2\pi$, it will follow from Remark 9.1 that

$$\int_{\Gamma \backslash G} |\langle \mathfrak{d}_{even}^{g}, u \rangle|^2 \, dg = \frac{4\pi}{\sqrt{27}} \| u \|^2 \tag{4.1}$$

if u is an even function while, if u is odd,

$$\int_{\Gamma \backslash G} |\langle \mathfrak{d}_{odd}^{g}, u \rangle|^2 \, dg = \frac{2\pi}{3} \| u \|^2. \tag{4.2}$$

This puts forward the roles of the two families $(\mathfrak{d}_{even}^{\tilde{g}})$ and $(\mathfrak{d}_{odd}^{\tilde{g}})$ as families of coherent states in the sense given in the introduction. It also explains our interest in the computation of the associated Wigner functions.

The preceding two equations are also a particular case of the results of Section 12: starting from a number N, the product of 4 by a squarefree odd integer, and using the notation of Lemma 3.2 again,

$$\sum_{\rho \in R_N} \int_{\Gamma \backslash G} |\langle \varpi_\rho^{g}, u \rangle|^2 \, dg = \frac{2\pi}{3}\, N^{-\frac{1}{2}}\, \phi_{\mathcal{E}}(N) \, \| u \|_{L^2(\mathbb{R})}^2, \tag{4.3}$$

where $\phi_{\mathcal{E}}$ is Euler's indicator function. As will be explained in the proof of Lemma 12.2, nothing analogous, however, could work with ∂_0 substituted for ∂_{even} or ∂_{odd}.

Equations (4.1), (4.2) and (4.3) have been displayed here for a good comprehension, but of course nothing depends on such results, to be proved only later.

Lemma 4.1. *Let N be the product of 4 by a squarefree odd integer, and keep the notation of Lemma 3.2 regarding Λ and R_N. For every pair (j, k) of integers mod N, set*

$$\mathfrak{D}^{jk}(x, \xi) = \sum_{m, n \in \mathbb{Z}} \delta(x - j - mN)\, \delta(\xi - k - nN), \qquad (4.4)$$

thus defining a distribution invariant under the linear action of matrices $\left(\begin{smallmatrix} a & b \\ c & d \end{smallmatrix}\right)$ in the principal congruence group [20] defined by the equation $\left(\begin{smallmatrix} a & b \\ c & d \end{smallmatrix}\right) = \pm\left(\begin{smallmatrix} 1 & 0 \\ 0 & 1 \end{smallmatrix}\right)$ mod N. For every pair (ρ_1, ρ_2) in R_N, the Wigner function $W(\varpi_{\rho_1}, \varpi_{\rho_2})$ is a linear combination of the distributions $N^{i\pi\mathcal{E}}\,\mathfrak{D}^{jk}$.

Proof. It is immediate to verify that, given any matrix $\left(\begin{smallmatrix} a & b \\ c & d \end{smallmatrix}\right) \in SL(2, \mathbb{Z})$, one has

$$\mathfrak{D}^{jk}(ax + b\xi, cx + d\xi) = \mathfrak{D}^{dj-bk, -cj+ak}(x, \xi), \qquad (4.5)$$

which justifies the first assertion. The computation of Wigner functions starts from the equation

$$W(\delta_a, \delta_b)(x, \xi) = \delta\left(x - \frac{a+b}{2}\right) e^{2i\pi(a-b)\xi}, \qquad (4.6)$$

which can be checked by verifying that

$$(\text{Op}(W(\delta_a, \delta_b))\, u)(x) = u(a)\, \delta(x - b) \qquad (4.7)$$

for every $u \in \mathcal{S}(\mathbb{R})$.

Using (3.9) and (4.6), one obtains, for any pair ρ_1, ρ_2 in R_N,

$$W(\varpi_{\rho_1}, \varpi_{\rho_2})(x, \xi) = \sum_{\mu_1, \mu_2 \in \Lambda} \chi(\mu_1 \mu_2) \sum_{\ell_1, \ell_2 \in \mathbb{Z}} \delta\left(x - \frac{N(\ell_1 + \ell_2) + \langle \rho, \mu \rangle}{2\sqrt{N}}\right)$$

$$\exp\left(2i\pi\xi\, \frac{N(\ell_1 - \ell_2) + \langle J\rho, \mu \rangle}{\sqrt{N}}\right), \qquad (4.8)$$

where we have set

$$\langle \rho, \mu \rangle = \rho_1 \mu_1 + \rho_2 \mu_2, \qquad \langle J\rho, \mu \rangle = \rho_1 \mu_1 - \rho_2 \mu_2. \qquad (4.9)$$

Setting $(\ell_1 + \ell_2, \ell_1 - \ell_2) = (2m + \eta, 2k + \eta)$ with $m \in \mathbb{Z}$, $k \in \mathbb{Z}$, $\eta = 0$ or 1, one obtains, since

$$\sum_{k \in \mathbb{Z}} e^{4i\pi k\sqrt{N}\xi} = \frac{1}{2\sqrt{N}} \sum_{k \in \mathbb{Z}} \delta\left(\xi - \frac{k}{2\sqrt{N}}\right) \qquad (4.10)$$

by Poisson's formula,

$$W(\varpi_{\rho_1}, \varpi_{\rho_2})(x, \xi) = \frac{1}{2\sqrt{N}} \sum_{\eta=0,1} \sum_{\mu_1, \mu_2 \in \Lambda} \chi(\mu_1 \mu_2)$$

$$\times \sum_{m,k \in \mathbb{Z}} \delta\left(x - \frac{N(2m+\eta) + \langle \rho, \mu \rangle}{2\sqrt{N}}\right) \delta\left(\xi - \frac{k}{2\sqrt{N}}\right) \exp\left(2i\pi\xi \frac{N\eta + \langle J\rho, \mu \rangle}{\sqrt{N}}\right),$$

$$(4.11)$$

or

$$W(\varpi_{\rho_1}, \varpi_{\rho_2})(x, \xi) = \frac{1}{2\sqrt{N}} \sum_{\mu_1, \mu_2 \in \Lambda} \chi(\mu_1 \mu_2)$$

$$\times \left[\sum_{m,k \in \mathbb{Z}} \delta\left(x - \frac{N(2m) + \langle \rho, \mu \rangle}{2\sqrt{N}}\right) \delta\left(\xi - \frac{k}{2\sqrt{N}}\right) \exp\left(\frac{i\pi k \langle J\rho, \mu \rangle}{N}\right) \right.$$

$$\left. + (-1)^k \sum_{m,k \in \mathbb{Z}} \delta\left(x - \frac{N(2m+1) + \langle \rho, \mu \rangle}{2\sqrt{N}}\right) \delta\left(\xi - \frac{k}{2\sqrt{N}}\right) \exp\left(\frac{i\pi k \langle J\rho, \mu \rangle}{N}\right) \right].$$

$$(4.12)$$

Finally,

$$W(\varpi_{\rho_1}, \varpi_{\rho_2})(x, \xi) = \frac{1}{2\sqrt{N}} \sum_{j, k \in \mathbb{Z}} B_{jk}(\rho_1, \rho_2) \delta\left(x - \frac{j}{2\sqrt{N}}\right) \delta\left(\xi - \frac{k}{2\sqrt{N}}\right)$$

$$(4.13)$$

with

$$B_{jk}(\rho_1, \rho_2) = \sum_{\mu_1, \mu_2 \in \Lambda} \chi(\mu_1 \mu_2) \exp\left(\frac{i\pi k \langle J\rho, \mu \rangle}{N}\right)$$

$$\times \left[\text{char}\left(j \equiv \langle \rho, \mu \rangle \bmod 2N\right) + (-1)^k \text{char}\left(j \equiv \langle \rho, \mu \rangle + N \bmod 2N\right)\right]. \quad (4.14)$$

It is not immediately apparent that this sum, as it should, only depends on ρ_1, ρ_2 as defined mod N, not $2N$. However, since $\exp\left(i\pi k \frac{\langle \rho, \mu \rangle - j}{N}\right)$ coincides with 1 (*resp.* $(-1)^k$) when $j \equiv \langle \rho, \mu \rangle \bmod 2N$ (*resp.* $j \equiv \langle \rho, \mu \rangle + N \bmod 2N$), one may write

$$B_{jk}(\rho_1, \rho_2) = e^{-\frac{i\pi jk}{N}} \sum_{\substack{\mu_1, \mu_2 \in \Lambda \\ \langle \rho, \mu \rangle \equiv j \bmod N}} \chi(\mu_1 \mu_2) \exp\left(2i\pi k \frac{\rho_1 \mu_1}{N}\right), \quad (4.15)$$

from which the desired fact is obvious. Moreover, since N is even, $\rho_1, \mu_1, \rho_2, \mu_2 \in (\mathbb{Z}/N\mathbb{Z})^\times$ are all relatively prime to 2, so that $\langle \rho, \mu \rangle$ is even: as a consequence,

$B_{jk}(\rho_1, \rho_2) = 0$ unless j is even. From the covariance of the Weyl calculus, one obtains

$$W(\mathcal{F}\varpi_{\rho_1}, \mathcal{F}\varpi_{\rho_2})(x, \xi) = \frac{1}{2\sqrt{N}} \sum_{j,k \in \mathbb{Z}} B_{-k, j}(\rho_1, \rho_2)\delta\left(x - \frac{j}{2\sqrt{N}}\right)\delta\left(\xi - \frac{k}{2\sqrt{N}}\right):$$

(4.16)

on the other hand, from Lemma 3.2, one has

$$W(\mathcal{F}\varpi_{\rho_1}, \mathcal{F}\varpi_{\rho_2}) = \sum_{\sigma_1, \sigma_2} K(\rho_1, \sigma_1)\, W(\varpi_{\sigma_1}, \varpi_{\sigma_2})\, \overline{K}(\rho_2, \sigma_2),$$

(4.17)

so that

$$B_{-k, j}(\rho_1, \rho_2) = \sum_{\sigma_1, \sigma_2} K(\rho_1, \sigma_1)\, B_{jk}(\sigma_1, \sigma_2)\, \overline{K}(\rho_2, \sigma_2).$$

(4.18)

It follows that B_{jk} can be non-identically zero, as a function on $R_N \times R_N$, only if both j and k are even. One may thus rewrite (4.13) as

$$W(\varpi_{\rho_1}, \varpi_{\rho_2})(x, \xi) = \frac{1}{\sqrt{N}} \sum_{j,k \in \mathbb{Z}} \Gamma_{jk}(\rho_1, \rho_2)\, \delta\left(x - \frac{j}{\sqrt{N}}\right)\delta\left(\xi - \frac{k}{\sqrt{N}}\right),$$

(4.19)

with

$$\Gamma_{jk}(\rho_1, \rho_2) = \frac{1}{2}\, B_{2j, 2k}(\rho_1, \rho_2)$$

$$= \frac{1}{2} \sum_{\substack{\mu_1, \mu_2 \in \Lambda \\ \langle \rho, \mu \rangle \equiv 2j \bmod N}} \chi(\mu_1 \mu_2)\, \exp\left(2i\pi k\, \frac{\langle J\rho, \mu \rangle}{N}\right).$$

(4.20)

One may note that $\Gamma_{jk}(\rho_1, \rho_2)$ only depends on the pair $(j, k) \bmod \frac{N}{2}$. Also, one has the relations, valid for every $\nu \in \Lambda$,

$$\Gamma_{jk}(\rho_1, \nu\rho_2) = \Gamma_{jk}(\nu\rho_1, \rho_2) = \chi(\nu)\, \Gamma_{jk}(\rho_1, \rho_2):$$

(4.21)

hence, up to the multiplication by ± 1, $\Gamma_{jk}(\rho_1, \rho_2)$ only depends on the pair (ρ_1^2, ρ_2^2).

Extending the operator $t^{2i\pi\mathcal{E}}$, as defined in (2.7) as an operator on functions in the plane, to distributions $\mathfrak{S}$, i.e., by means of the equation

$$\langle t^{1+2i\pi\mathcal{E}}\, \mathfrak{S}, h \rangle = \langle \mathfrak{S}, (x, \xi) \mapsto h(t^{-1}x, t^{-1}\xi) \rangle,$$

(4.22)

one has, for $t > 0$,

$$t^{-1-2i\pi\mathcal{E}}\, (\delta(x - a)\, \delta(\xi - b)) = \delta(x - ta)\, \delta(\xi - tb),$$

(4.23)

and one can rewrite (4.19) as

$$W(\varpi_{\rho_1}, \varpi_{\rho_2}) = N^{i\pi\mathcal{E}}\, \mathfrak{S}_{\rho_1, \rho_2}$$

(4.24)

with

$$\mathfrak{S}_{\rho_1,\,\rho_2}(x,\,\xi) = \sum_{j,k\in\mathbb{Z}} \Gamma_{jk}(\rho_1,\,\rho_2)\,\delta(x-j)\,\delta(\xi-k) \tag{4.25}$$

$$= \sum_{j,k\in\mathbb{Z}/N\mathbb{Z}} \Gamma_{jk}(\rho_1,\,\rho_2)\,\mathfrak{D}^{jk}(x,\,\xi). \qquad\qquad \square$$

We now localize (in the arithmetic sense) the computation of the coefficients $\Gamma_{jk}(\rho_1,\,\rho_2)$, setting to begin with, whenever $p|N$,

$$\Lambda_p = \{\mu \bmod p\colon \mu^2 \equiv 1 \bmod p\} \text{ if } p \geq 3, \qquad \Lambda_2 = (\mathbb{Z}/4\mathbb{Z})^{\times}. \tag{4.26}$$

The canonical isomorphism $(\mathbb{Z}/N\mathbb{Z})^{\times} \sim \prod_{p\in S}(\mathbb{Z}/p^{\alpha_p}\mathbb{Z})^{\times}$ restricts as an isomorphism $\Lambda \sim \prod_{p\in S}\Lambda_p$; also, choosing for every $p \in S$ a set R_p of representatives of $(\mathbb{Z}/p^{\alpha_p}\mathbb{Z})^{\times} \bmod \Lambda_p$ (so that $R_p = \{1\}$ if $p = 2$, while R_p has $\frac{p-1}{2}$ elements if $p \geq 3$), the preceding isomorphism provides a bijection between the product $\prod_{p\in S} R_p$ and a set R_N (as introduced in Lemma 3.2) of representatives of $(\mathbb{Z}/N\mathbb{Z})^{\times} \bmod \Lambda$.

We now reconsider equation (4.20), setting according to what just precedes

$$\rho_1 = (\rho_1^{(p)})_{p\in S}, \quad \rho_2 = (\rho_2^{(p)})_{p\in S}, \quad \mu_1 = (\mu_1^{(p)})_{p\in S}, \quad \mu_2 = (\mu_2^{(p)})_{p\in S}: \tag{4.27}$$

also, j and k can be characterized by sets $(j_p)_{p\in S}$ and $(k_p)_{p\in S}$ respectively; one may assume that j_p, or k_p, is only given mod p for every $p \in S$, including the case when $p = 2$, since $\Gamma_{jk}(\rho_1,\,\rho_2)$ only depends on the pair (j,k) mod $\frac{N}{2}$. Denoting as χ_p the unique non-trivial character of Λ_p, one has $\chi(\mu_1) = \prod_{p\in S}\chi_p(\mu_1^{(p)})$ and the similar decomposition for $\chi(\mu_2)$. On the other hand, recall that M is defined by the equation $\frac{M}{N} = \sum_{p\in S}\frac{1}{p^{\alpha_p}}$: it is an integer relatively prime to N, and we denote as $\overline{M}$ any number the class of which in the group $(\mathbb{Z}/N\mathbb{Z})^{\times}$ is the inverse of that of M.

One has

$$\exp\left(2i\pi\frac{k\,\langle J\rho,\,\mu\rangle}{N}\right) = \exp\left(2i\pi\overline{M}\,k\sum_{p\in S}\frac{\langle J\rho,\,\mu\rangle}{p^{\alpha_p}}\right)$$

$$= \prod_{p\in S}\exp\left(2i\pi\overline{M}\,\frac{k_p\,\langle J\rho^{(p)},\,\mu^{(p)}\rangle}{p^{\alpha_p}}\right). \tag{4.28}$$

Hence,

$$\Gamma_{jk}(\rho_1,\,\rho_2) = \prod_{p\in S}\Gamma_{jk}^{(p)}(\rho_1,\,\rho_2), \tag{4.29}$$

where the local factors are defined by the equation, in which $|2|_2 = \frac{1}{2}$ and $|p|_2 = 1$ for $p \geq 3$,

$$\Gamma_{jk}^{(p)}(\rho_1, \rho_2) = |p|_2 \sum_{\substack{\mu_1, \mu_2 \in \Lambda_p \\ \langle \rho, \mu \rangle \equiv 2j \\ \bmod p^{\alpha_p}}} \chi_p(\mu_1)\, \chi_p(\mu_2) \exp\left(\frac{2i\pi \overline{M} k}{p^{\alpha_p}} \langle J\rho, \mu \rangle \right) : \qquad (4.30)$$

in the local factor at p, ρ_1 (*resp.* ρ_2) can of course be interpreted as a class mod p^{α_p}, while j (*resp.* k) can be interpreted as a class mod p. Note that, because of the occurrence of $\overline{M}$, the local factor at p depends on N (at least on M_p, since $M_p \equiv M \bmod p^{\alpha_p}$), not only on p.

Set $N' = \frac{N}{2}$. The function $\Psi^{(N)} : \mathbb{Z}/N'\mathbb{Z} \times \mathbb{Z}/N'\mathbb{Z} \to \mathcal{L}(\mathbb{C}^{R_N}, \mathbb{C}^{R_N})$ such that

$$\Psi^{(N)}(j, k) = (\Gamma_{jk}(\rho_1, \rho_2))_{\rho_1, \rho_2 \in R_N} \qquad (4.31)$$

is thus the "tensor" product of the local functions $\Psi^{(p^{\alpha_p})}$ at points $p \in S$, once $\mathbb{Z}/N'\mathbb{Z}$ has been identified with the product $\prod_{p \in S} \mathbb{Z}/p\mathbb{Z}$, and the space $\mathcal{L}(\mathbb{C}^{R_N}, \mathbb{C}^{R_N})$ has been identified with the tensor (Kronecker) product of the local spaces of matrices $\mathcal{L}(\mathbb{C}^{R_p}, \mathbb{C}^{R_p})$. We shall come back to a study of this set of coefficients in general in the last section.

In the case when $p = 2$ or 3, R_p reduces to one element, which we choose of course as the unit element of $(\mathbb{Z}/p^{\alpha_p}\mathbb{Z})^\times$, and we abbreviate $\Gamma_{jk}^{(p)}(\rho_1, \rho_2)$ as $\Gamma_{jk}^{(p)}$. One has in this case

$$\Gamma_{jk}^{(p)} = |p|_2 \sum_{\substack{\mu_1, \mu_2 \in \Lambda_p \\ \mu_1 + \mu_2 \equiv 2j \\ \bmod p^{\alpha_p}}} \chi_p(\mu_1)\, \chi_p(\mu_2) \exp\left(\frac{2i\pi \overline{M} k}{p^{\alpha_p}} (\mu_1 - \mu_2) \right). \qquad (4.32)$$

If $p = 2$, one has $\chi_2(\mu_1)\,\chi_2(\mu_2) = -1$ if j is even, and $\chi_2(\mu_1)\,\chi_2(\mu_2) = 1$ if j is odd: it follows, using also the fact that $\overline{M}$ is in any case an odd number, that $\Gamma_{jk}^{(2)} = -1$ if j and k are even, 1 in all other cases. Hence the scalar function $\Psi^{(4)}$ is encoded by the matrix

$$\Psi^{(4)} = \begin{pmatrix} -1 & 1 \\ 1 & 1 \end{pmatrix}. \qquad (4.33)$$

If $p = 3$, one has again

$$\chi_3(\mu_1)\,\chi_3(\mu_2) = -1 \text{ if } j \equiv 0 \quad \bmod 3, \quad \text{and} \quad \chi_3(\mu_1)\,\chi_3\mu_2) = 1 \text{ if } j \not\equiv 0 \quad \bmod 3.$$

Note that when $j \equiv 0 \bmod 3$, there are two choices for the pair (μ_1, μ_2), to wit $(1, 2)$ and $(2, 1)$; on the contrary, when $j \equiv 1$ (*resp.* 2) mod 3, one must take $(\mu_1, \mu_2) = (1, 1)$ (*resp.* $(2, 2)$). This yields (independently of the value of $\overline{M} = 1$

or 2) $\Gamma_{jk}^{(p)} = -2$ when j and k are divisible by 3, 1 in all other cases. Hence, the scalar function $\Psi^{(3)}$ is encoded by the matrix

$$\Psi^{(3)} = \begin{pmatrix} -2 & 1 & 1 \\ 1 & 1 & 1 \\ 1 & 1 & 1 \end{pmatrix}. \tag{4.34}$$

Theorem 4.2. *Recall that* $\eth_0(x) = \sum_{j \in \mathbb{Z}} \delta(x - j)$, *and that* $\eth_{even}$ *and* $\eth_{odd}$ *have been defined in Theorem 3.1. Setting*

$$\mathfrak{D}_0(x, \xi) = \sum_{j,\,k \in \mathbb{Z}} \delta(x - j)\, \delta(\xi - k)\,,$$

$$\mathfrak{D}_1(x, \xi) = \frac{1}{2} \sum_{j,\,k \in \mathbb{Z}} e^{i\pi jk}\, \delta\left(x - \frac{j}{2}\right) \delta\left(\xi - \frac{k}{2}\right), \tag{4.35}$$

one has

$$W(\eth_0, \eth_0) = \mathfrak{D}_1\,,$$
$$W(\eth_{even}, \eth_{even}) = 12^{i\pi\,\mathcal{E}} \left[1 - 2^{-2i\pi\,\mathcal{E}}\right] \left[1 - 3^{-2i\pi\,\mathcal{E}}\right] \mathfrak{D}_0\,,$$
$$W(\eth_{odd}, \eth_{odd}) = 4^{i\pi\,\mathcal{E}} \left[1 - 2^{-2i\pi\mathcal{E}}\right] \mathfrak{D}_0\,. \tag{4.36}$$

Proof. The first equation is a simple consequence of (4.6), as one can write

$$W(\eth_0, \eth_0)(x, \xi) = \sum_{m,n \in \mathbb{Z}} \delta\left(x - \frac{m+n}{2}\right) e^{2i\pi(m-n)\xi}$$

$$= \sum_{k\ \text{even}} e^{2i\pi k\xi} \sum_{j \in \mathbb{Z}} \delta(x - j) + \sum_{k\ \text{odd}} e^{2i\pi k\xi} \sum_{j \in \mathbb{Z}} \delta\left(x - j - \frac{1}{2}\right), \tag{4.37}$$

and then apply Poisson's formula to transform the sum of exponentials.

To compute $W(\eth_{even}, \eth_{even})$, we must apply (4.19), where the coefficients Γ_{jk} are determined as the entries of the matrix

$$A = \begin{pmatrix} 2 & 1 & -1 & -2 & -1 & 1 \\ 1 & 1 & 1 & 1 & 1 & 1 \\ -1 & 1 & -1 & 1 & -1 & 1 \\ -2 & 1 & 1 & -2 & 1 & 1 \\ -1 & 1 & -1 & 1 & -1 & 1 \\ 1 & 1 & 1 & 1 & 1 & 1 \end{pmatrix}, \tag{4.38}$$

the lines and columns of which are parametrized by the elements from 0 to 5 of $\mathbb{Z}/6\mathbb{Z}$. According to (4.29), the matrix has been obtained by the following recipe: first, change (4.33) and (4.34) into 6×6-matrices by periodicity mod 2 or 3; next, multiply the corresponding entries of the resulting matrices.

One has $A = A_1 - 2A_2 - 3A_3 + 6A_6$ if the new 6×6-matrices are defined as follows: every entry of a^i_{jk} of A_i is 0 except when j and k are both divisible by i, in which case it is 1: recall that j, k start from 0, not 1,. One may thus write

$$W^{\text{even,even}}(x, \xi)$$
$$= \frac{1}{\sqrt{12}} \sum_{j,k \in \mathbb{Z}} \left[\delta\left(x - \frac{j}{\sqrt{12}}\right) \delta\left(\xi - \frac{k}{\sqrt{12}}\right) - 2\delta\left(x - \frac{2j}{\sqrt{12}}\right) \delta\left(\xi - \frac{2k}{\sqrt{12}}\right) \right.$$
$$\left. - 3\delta\left(x - \frac{3j}{\sqrt{12}}\right) \delta\left(\xi - \frac{3k}{\sqrt{12}}\right) + 6\delta\left(x - \frac{6j}{\sqrt{12}}\right) \delta\left(\xi - \frac{6k}{\sqrt{12}}\right) \right] \quad (4.39)$$

or, using (4.23) again,

$$W^{\text{even,even}} = 12^{i\pi\mathcal{E}} \left[1 - 2^{-2i\pi\mathcal{E}} - 3^{-2i\pi\mathcal{E}} + 6^{-2i\pi\mathcal{E}} \right] \mathfrak{D}_0, \quad (4.40)$$

which is the first equation (4.36).

Directly from (4.33) and, again, from (4.19), one obtains

$$W^{\text{odd,odd}}(x, \xi) = \frac{1}{2} \sum_{m,k \in \mathbb{Z}} \left[\delta\left(x - m + \frac{1}{2}\right) + (-1)^{k+1} \delta(x - m) \right] \delta\left(\xi - \frac{k}{2}\right)$$
$$= \frac{1}{2} \sum_{j,k \in \mathbb{Z}} (-1)^{(j-1)(k-1)} \delta\left(x - \frac{j}{2}\right) \delta\left(\xi - \frac{k}{2}\right), \quad (4.41)$$

from which one finds the second equation (4.36). $\square$

Some verifications are useful here: first, both $W^{\text{even,even}}$ and $W^{\text{odd,odd}}$ are obviously even and $SL(2,\mathbb{Z})$-automorphic. Next, using the equation $\mathcal{G} = 2^{2i\pi\mathcal{E}} \mathcal{F}$, so that $\mathcal{G}\mathfrak{D}_0 = 2^{2i\pi\mathcal{E}} \mathfrak{D}_0$, and the fact that the conjugation by $\mathcal{G}$ changes $\mathcal{E}$ to its negative, one verifies that, as was to be expected as a consequence of (2.9), $W^{\text{even,even}}$ is $\mathcal{G}$-invariant and $W^{\text{odd,odd}}$ changes to its negative under $\mathcal{G}$. The identity $W(\mathfrak{d}_{\text{odd}}, \mathfrak{d}_{\text{odd}})(x, \xi) = W(\mathfrak{d}_0, \mathfrak{d}_0)(x - \frac{1}{2}, \xi - \frac{1}{2})$, on the other hand, can be explained by the covariance of the Weyl calculus under the Heisenberg representation, since $\mathfrak{d}_{\text{odd}}(x) = \mathfrak{d}_0(x - \frac{1}{2}) e^{i\pi(x - \frac{1}{2})}$.

We now consider the transforms of $\mathfrak{d}_{\text{even}} = \mathfrak{d}^{(12)}$ and $\mathfrak{d}_{\text{odd}} = \mathfrak{d}^{(4)}$ under metaplectic transformations lying above matrices in G with *rational* coefficients. The homogeneous space $SL(2,\mathbb{Q})/SL(2,\mathbb{Z})$ can be identified with the set of rational symplectic lattices Λ in $\mathbb{R}^2$: this is by definition a lattice with a basis consisting of vectors in $\mathbb{Q}^2$, with the additional property that a fundamental parallelogram has area (in absolute value) 1: the lattice corresponding to the class $g.SL(2,\mathbb{Z})$ with $g = \left(\begin{smallmatrix} a & b \\ c & d \end{smallmatrix}\right)$ is the one generated by the vectors $\left(\begin{smallmatrix} a \\ c \end{smallmatrix}\right)$ and $\left(\begin{smallmatrix} b \\ d \end{smallmatrix}\right)$. We now show that this set also parametrizes – up to some phase factors – the set of transforms of $\mathfrak{d}^{(12)} = \mathfrak{d}_{\text{even}}$ or $\mathfrak{d}^{(4)} = \mathfrak{d}_{\text{odd}}$ under all metaplectic transformations lying above matrices in $SL(2,\mathbb{Q})$.

Proposition 4.3. *Given a rational symplectic lattice* Λ, *one can find a matrix* $\begin{pmatrix} a & 0 \\ c & a^{-1} \end{pmatrix} \in SL(2,\mathbb{Q})$ *with* $a > 0$ *such that* Λ *is generated by the vectors* $\begin{pmatrix} a \\ c \end{pmatrix}$ *and* $\begin{pmatrix} 0 \\ a^{-1} \end{pmatrix}$. *The number* $a \in \mathbb{Q}^+$ *is unique, and the number* $ac \in \mathbb{Q}$ *is unique up to the addition of an arbitrary integer: to define it without ambiguity, we make the arbitrary choice that it is minimum among the non-negative possible choices. With* $N = 12$ *or* 4, *and* $\chi^{(N)}$ *as defined in Theorem 3.1, set*

$$\mathfrak{d}_\Lambda^{(N)}(x) = a^{\frac{1}{2}} \sum_{m \in \mathbb{Z}} \chi^{(N)}(m) \, e^{\frac{i\pi acm^2}{N}} \, \delta\left(x - \frac{am}{\sqrt{N}}\right). \tag{4.42}$$

The set of transforms of $\mathfrak{d}^{(N)}$ *under all metaplectic transformations lying above matrices* g *in the group* $SL(2,\mathbb{Q})$ *such that the class* $g \cdot SL(2,\mathbb{Z})$ *defines the rational symplectic lattice* Λ *is exactly the set of distributions*

$$x \mapsto e^{i\theta} \, \mathfrak{d}_\Lambda^{(N)} \tag{4.43}$$

where $e^{i\theta}$ *is an arbitrary* $(2N)$*th root of unity. Set*

$$\langle \mathfrak{D}_0^\Lambda, h \rangle = \sum_{(\mu,\nu) \in \Lambda} h(\mu,\nu), \qquad h \in \mathcal{S}(\mathbb{R}^2). \tag{4.44}$$

Then, one has the following two formulas, making Wigner functions in the Weyl calculus explicit:

$$W(\mathfrak{d}_\Lambda^{(12)}, \mathfrak{d}_\Lambda^{(12)}) = 12^{i\pi\,\mathcal{E}} \left[1 - 2^{-2i\pi\,\mathcal{E}} \right] \left[1 - 3^{-2i\pi\,\mathcal{E}} \right] \mathfrak{D}_0^\Lambda,$$
$$W(\mathfrak{d}_\Lambda^{(4)}, \mathfrak{d}_\Lambda^{(4)}) = 4^{i\pi\,\mathcal{E}} \left[1 - 2^{-2i\pi\,\mathcal{E}} \right] \mathfrak{D}_0^\Lambda. \tag{4.45}$$

Proof. The transforms of $\mathfrak{d}^{(N)}$ under arbitrary metaplectic transformations lying above matrices in $SL(2,\mathbb{Z})$ are exactly the distributions $e^{i\theta} \, \mathfrak{d}^{(N)}$ with $e^{i\theta}$ as indicated, as it follows from Theorem 3.1 and from the computation of the matrices (in this case, complex numbers) T and K in (3.19). If $\begin{pmatrix} a & b \\ c & d \end{pmatrix} \in SL(2,\mathbb{Q})$, one has $-Nb\begin{pmatrix} a \\ c \end{pmatrix} + Na\begin{pmatrix} b \\ d \end{pmatrix} = \begin{pmatrix} 0 \\ N \end{pmatrix}$, and N can be chosen so that Na and Nb are integers. Hence, the lattice $\Lambda \subset \mathbb{Q}^2$ generated by $\begin{pmatrix} a \\ c \end{pmatrix}$ and $\begin{pmatrix} b \\ d \end{pmatrix}$ intersects $0 \times \mathbb{Q}$ along a non-zero subgroup Λ_1. Choosing the generator $\begin{pmatrix} 0 \\ y_2 \end{pmatrix}$ of Λ_1 with $y_2 > 0$ and a vector $\begin{pmatrix} x_1 \\ x_2 \end{pmatrix}$ with $x_1 > 0$, the class of which mod Λ_1 generates Λ/Λ_1, one sees that Λ is also generated by $\begin{pmatrix} x_1 \\ x_2 \end{pmatrix}$ and $\begin{pmatrix} 0 \\ y_2 \end{pmatrix}$, so that $\begin{pmatrix} a & b \\ c & d \end{pmatrix} = \begin{pmatrix} x_1 & 0 \\ x_2 & y_2 \end{pmatrix} \gamma$ for some $\gamma \in SL(2,\mathbb{Z})$. Note that y_2 and $x_1 = y_2^{-1}$ are unique, but that x_2 is only unique up to the addition of a multiple of y_2. Changing notation, we are left with having to compute only $\mathrm{Met}(\tilde{g}) \, \mathfrak{d}^{(N)}$ for elements $\tilde{g}$ of the metaplectic group lying above matrices of the kind

$$\begin{pmatrix} a & 0 \\ c & a^{-1} \end{pmatrix} = \begin{pmatrix} 1 & 0 \\ \frac{c}{a} & 1 \end{pmatrix} \begin{pmatrix} a & 0 \\ 0 & a^{-1} \end{pmatrix}, \tag{4.46}$$

with $a > 0$ and $ac \in \mathbb{Q}$ known up to the addition of an arbitrary integer. The table given at the very beginning of Section 2 gives the answer: one finds a distribution

$$\pm a^{-\frac{1}{2}} e^{\frac{i\pi c x^2}{a}} \mathfrak{d}(a^{-1}x) = \pm a^{\frac{1}{2}} e^{\frac{i\pi c x^2}{a}} \sum_{m \in \mathbb{Z}} \chi^{(N)}(m) \, \delta \left(x - \frac{am}{\sqrt{N}} \right)$$

$$= \pm a^{\frac{1}{2}} e^{\frac{i\pi a c m^2}{N}} \sum_{m \in \mathbb{Z}} \chi^{(N)}(m) \, \delta \left(x - \frac{am}{\sqrt{N}} \right). \qquad (4.47)$$

It remains to be observed that if one changes ac to $ac + 1$, so that $e^{\frac{i\pi a c m^2}{N}}$ is multiplied by $e^{\frac{i\pi m^2}{N}}$, this has the same effect as multiplying the whole distribution by $e^{\frac{i\pi}{N}}$ since, as part of the construction in Section 3, one has $m^2 \equiv 1 \mod 2N$ whenever $\chi^{(N)}(m) \neq 0$.

The last point is a consequence of Theorem 4.2 and of the covariance property of the Weyl calculus. Indeed, if $g \in SL(2, \mathbb{R})$, one has

$$W(\mathfrak{d}_\Lambda^{(N)}, \mathfrak{d}_\Lambda^{(N)}) = W(\mathfrak{d}^{(N)}, \mathfrak{d}^{(N)}) \circ g^{-1} \qquad (4.48)$$

if Λ is associated to the class $g \, . \, SL(2, \mathbb{Z})$. Next, one has

$$\langle \mathfrak{D}_0 \circ g^{-1}, h \rangle = \langle \mathfrak{D}_0, h \circ g \rangle$$

$$= \sum_{j,k \in \mathbb{Z}} (h \circ g)(j, k) = \sum_{(\mu,\nu) \in \Lambda} h(\mu, \nu). \qquad (4.49)$$

Finally, the operator $2i\pi\mathcal{E}$ commutes with the linear action of $SL(2, \mathbb{R})$ in $\mathbb{R}^2$. $\square$

Obviously, the formula (4.42) does not lend itself to a generalization valid for a matrix with arbitrary real coefficients. An adelic point of view could be more appropriate at this point, from the point of view of arithmetic, since it is clearly the Archimedean completion which does not seem to fit too well with the structure of the formula. However, the transform $\text{Met}(\tilde{g}) \, \mathfrak{d}^{(N)}$, for $g \in \widetilde{G}$ (the metaplectic group) still makes sense as a tempered distribution. We now give, in the odd case (*i.e.*, when $N = 4$) an explicit – in an analyst's sense – description of it. Recall from (2.12) the definition of the functions u_z^1.

Proposition 4.4. *Let* $g = \left(\begin{smallmatrix} a & b \\ c & d \end{smallmatrix} \right) \in G = SL(2, \mathbb{R})$, *with* $b > 0$, *be given, and let* $\tilde{g} \in \widetilde{G}$ *be the end point of a path, in* $\widetilde{G}$, *originating at the identity and covering a path, in* G, *consisting of matrices with a positive upper right entry. Then one has, in the weak sense against* $\mathcal{S}_{\text{odd}}(\mathbb{R})$, *the integral formula*

$$\text{Met}(\tilde{g}) \, \mathfrak{d}^{(4)} = \frac{1}{8\pi} \int_\Pi \Phi(z) \left(\frac{a - b\bar{z}}{|a - b\bar{z}|} \right)^{-\frac{3}{2}} u_{\frac{bz-a}{dz-c}}^1 \, d\mu(z), \qquad (4.50)$$

where the fractional power is associated to the determination of logarithms, in the upper half-plane, with an imaginary part in $]0, \pi[$, and the function Φ is defined as

$$\Phi(z) = 2^{\frac{1}{4}} \pi^{\frac{1}{2}} (\operatorname{Im} z)^{\frac{3}{4}} \sum_{m \in \mathbb{Z}} \chi^{(4)}(m) \cdot m \, e^{\frac{i \pi z m^2}{4}} . \tag{4.51}$$

Proof. Starting from the equation (2.13) and using the isometric operator $\mathrm{Sq}_{\mathrm{odd}}$ from $H_{\frac{3}{2}}$ to $L^2_{\mathrm{odd}}(\mathbb{R})$ (intertwining the representation $\mathcal{D}_{\frac{3}{2}}$ with the restriction to $L^2_{\mathrm{odd}}(\mathbb{R})$ of the metaplectic representation), one obtains

$$\mathrm{Met}(\tilde{g}) \, \mathfrak{d}^{(4)} = \frac{1}{8\pi} \int_{\Pi} (u_z^1 \mid \mathfrak{d}^{(4)}) \, \mathrm{Met}(\tilde{g}) \, u_z^1 \, d\mu(z)$$

$$= \frac{1}{8\pi} \int_{\Pi} (u_{-\frac{1}{z}}^1 \mid \mathfrak{d}^{(4)}) \, \mathrm{Met}(\tilde{g}) \, u_{-\frac{1}{z}}^1 \, d\mu(z). \tag{4.52}$$

Since

$$u_{-\frac{1}{z}}^1(t) = 2^{\frac{5}{4}} \pi^{\frac{1}{2}} (\operatorname{Im} z)^{\frac{3}{4}} t \, e^{-i\pi \bar{z} t^2}, \tag{4.53}$$

it is immediate to obtain, setting $\Phi(z) = (u_{-\frac{1}{z}}^1 \mid \mathfrak{d}^{(4)})$, the expression of $\Phi(z)$ indicated. On the other hand, using the decomposition

$$\begin{pmatrix} a & b \\ c & d \end{pmatrix} = \begin{pmatrix} 1 & 0 \\ \frac{d}{b} & 1 \end{pmatrix} \begin{pmatrix} b & 0 \\ 0 & \frac{1}{b} \end{pmatrix} \begin{pmatrix} 0 & 1 \\ -1 & 0 \end{pmatrix} \begin{pmatrix} 1 & 0 \\ \frac{a}{b} & 1 \end{pmatrix}, \tag{4.54}$$

one can see that

$$\mathrm{Met}(\tilde{g}) \, u_{-\frac{1}{z}}^1 = \left(\frac{a - b\bar{z}}{|a - b\bar{z}|} \right)^{-\frac{3}{2}} u_{\frac{bz-a}{dz-c}}^1 : \tag{4.55}$$

it suffices to apply the formulas given in the beginning of Section 2. The only minor headache concerns the phase factor in connection with the second matrix from the right: with an obvious notation, the element of the metaplectic group above this matrix must be interpreted as $\begin{pmatrix} \cos \frac{\pi}{2} & \sin \frac{\pi}{2} \\ -\sin \frac{\pi}{2} & \cos \frac{\pi}{2} \end{pmatrix}$, and the associated metaplectic transformation is $e^{-\frac{i\pi}{4}} \mathcal{F}$.

The proof is over. Let us remark, however, that one cannot do the same with the even distribution $\mathfrak{d}^{(12)}$: the reason is that there is no analogue of (2.13) in this case, since the representation $\mathcal{D}_{\frac{1}{2}}$ is not square integrable. However, one can still find an integral formula in the weak sense against functions in $\mathcal{S}_{\mathrm{even}}(\mathbb{R})$ which are orthogonal to u_i (u_z would do just as well for any z), writing every such function as the image of an odd function under the so-called creation operator $\pi^{\frac{1}{2}} \left(t - \frac{1}{2\pi} \frac{d}{dt} \right)$ and taking advantage of what has been done in the odd case. $\square$

Remark 4.1. In the beginning of this section, we have emphasized the roles of the distributions $\mathfrak{d}_{\mathrm{even}}^{\tilde{g}}$, $\mathfrak{d}_{\mathrm{odd}}^{\tilde{g}}$ or of the collection $(\varpi_\rho^{\tilde{g}})_{\rho \in R_N}$ as families of arithmetic coherent states for the metaplectic representation and for the arithmetic subgroup $SL(2, \mathbb{Z})$ of $SL(2, \mathbb{R})$. An analogous, easier to establish, and quite well-known fact, concerns the existence of a family of arithmetic coherent states for the Heisenberg representation, or the associated projective representation of the additive group $\mathbb{R}^2$ in $L^2(\mathbb{R})$, and the subgroup $\mathbb{Z}^2$ of $\mathbb{R}^2$, a fundamental domain of which is the standard unit square D in $\mathbb{R}^2$. It is the set of distributions $(\mathfrak{d}_{x,\xi})_{(x,\xi)\in\mathbb{R}^2}$ such that

$$\mathfrak{d}_{x,\xi}(t) = \sum_{m\in\mathbb{Z}} \delta(t - x - m)\, e^{2i\pi(t-\frac{x}{2})\xi} : \qquad (4.56)$$

up to the multiplication by some phase factor, $\mathfrak{d}_{x,\xi}$ only depends on the class of $(x,\xi) \bmod \mathbb{Z}^2$ so that, for any $u \in \mathcal{S}(\mathbb{R})$, the function $(\Theta u)(x,\xi) = (\mathfrak{d}_{x,\xi}\,|\,u)$ satisfies a "pseudo-periodicity" condition similar to that characteristic of theta-functions: of course, it is not holomorphic in any sense; the map Θ extends as an isometry from $L^2(\mathbb{R})$ onto the subspace of $L^2_{\mathrm{loc}}(\mathbb{R}^2)$ satisfying the above-mentioned pseudo-periodicity condition, such that $\int_D |(\Theta u)(x,\xi)|^2\, dx\, d\xi < \infty$. This fact [1], usually referred to as the Weil-Brezin theorem or, when applications to signal analysis are considered, the Weil-Brezin-Zak theorem, generalizes without modification to the higher-dimensional case.

The relationship of the developments centering around the Weil-Brezin transform and the ones in the present work is in more than one way analogous to the change of emphasis from the variable v to the variable τ, in theta-function theory: these include functions such as [7, p. 58]

$$\theta(v,\tau) = \frac{1}{i}\sum_{n\in\mathbb{Z}}(-1)^n\, q^{(n+\frac{1}{2})^2}\, e^{(2n+1)\pi i v}, \qquad q = e^{i\pi\tau}, \qquad (4.57)$$

and this change of emphasis, through the work, in particular, of Dedekind and Ramanujan, is the same as that which made the move from elliptic function theory to modular form theory a natural one [7, p. 136–140].

5 Matrix elements of operators from the Weyl calculus against arithmetic coherent states

Some knowledge of automorphic function theory is necessary in the present section: it will be recalled when needed. In the first part, leading to Theorem 5.1, we shall deal with the arithmetic group $\Gamma = SL(2,\mathbb{Z})$; in the second part, with the arithmetic group Γ_2 generated by the matrices $\left(\begin{smallmatrix} 0 & 1 \\ -1 & 0 \end{smallmatrix}\right)$ and $\left(\begin{smallmatrix} 1 & 0 \\ 2 & 1 \end{smallmatrix}\right)$. The domain D of the upper half-plane Π defined as $D = \Gamma\backslash\Pi = \{z \in \Pi : |z| > 1,\ |\mathrm{Re}\,z| < \frac{1}{2}\}$ in the first case, as $D = \Gamma_2\backslash\Pi = \{z \in \Pi : : |z| > 1,\ |\mathrm{Re}\,z| < 1\}$ in the second one, is a fundamental domain of the group Γ (*resp.* Γ_2): this means that every point of Π can be taken to some point of $\overline{D}$ under some transformation $z \mapsto g.z = \frac{az+b}{cz+d}$ with $\left(\begin{smallmatrix} a & b \\ c & d \end{smallmatrix}\right) \in \Gamma$ (*resp.* Γ_2), and that no two points of D can lie in the same orbit under such fractional-linear transformations from the group under consideration. The geometry of the first case is fully described in all textbooks on the subject, and that of the second one can be taken from the introduction of [18]; alternatively, one may remark (as will be shown later in this section) that Γ_2 is the conjugate, under some element of Γ, of the Hecke subgroup universally denoted as $\Gamma_0(2)$.

Because points $z = x + iy \in \Pi$ are essential objects here, we denote points of the phase space $\mathbb{R}^2$ as (q, p) rather than (x, ξ).

Let $\eth = \eth_{\text{even}}$ or $\eth_{\text{odd}}$, and let $\tilde{g}$ be a point of the metaplectic group lying above $g \in SL(2, \mathbb{R})$. Set $\eth^{\tilde{g}} = \text{Met}(\tilde{g}^{-1}) \eth$. Recall the equations (3.32), which make $\eth^{\tilde{g}}$ *almost* explicit in the case when $g = \left(\begin{smallmatrix} 0 & -1 \\ 1 & 0 \end{smallmatrix} \right)$ or $\left(\begin{smallmatrix} 1 & 0 \\ -1 & 1 \end{smallmatrix} \right)$. Given a symbol $h \in \mathcal{S}(\mathbb{R}^2)$, one may consider the function

$$
\begin{aligned}
(\eth^{\tilde{g}} \,|\, \text{Op}(h) \, \eth^{\tilde{g}}) &= \langle h, W(\eth^{\tilde{g}}, \eth^{\tilde{g}}) \rangle \\
&= \langle h, W(\eth, \eth) \circ g \rangle \\
&= \langle h \circ g^{-1}, W(\eth, \eth) \rangle,
\end{aligned}
\tag{5.1}
$$

which does not change if $\eth^{\tilde{g}}$ is multiplied by a complex number with absolute value 1, hence depends only on g, not $\tilde{g}$. Assume now that h is radial, $h(q, p) = H(q^2 + p^2)$: then, the result only depends on the class gK with $K = SO(2)$, so it can be expressed in terms of the point $z = g.i = x + iy$ of the upper half-plane.

Theorem 5.1. *Let $h \in \mathcal{S}(\mathbb{R}^2)$ be a radial function, and set, for $\lambda \in \mathbb{R}$,*

$$
\psi(i\lambda) = \frac{1}{4\pi^2} \int_{\mathbb{R}^2} (q^2 + p^2)^{\frac{-1-i\lambda}{2}} \, h(q, p) \, dq \, dp.
\tag{5.2}
$$

The functions of $z = g.i$ defined by the equations

$$
\begin{aligned}
(\mathcal{A}h)_0(z) &= (\eth^{\tilde{g}}_{\text{even}} \,|\, \text{Op}(h) \, \eth^{\tilde{g}}_{\text{even}}), \\
(\mathcal{A}h)_1(z) &= (\eth^{\tilde{g}}_{\text{odd}} \,|\, \text{Op}(h) \, \eth^{\tilde{g}}_{\text{odd}})
\end{aligned}
\tag{5.3}
$$

are automorphic with respect to the full modular group Γ. Set, for $\text{Re } s > 1$,

$$
E^*(z, s) = \frac{1}{2} \pi^{-s} \Gamma(s) \sum_{|m|+|n| \neq 0} \left(\frac{|m - nz|^2}{y} \right)^{-s}
\tag{5.4}
$$

and denote in the same way the analytic continuation of this function for $s \neq 0, 1$. The spectral (Roelcke-Selberg) decomposition of the first of the two functions above is given, under the assumption that h is $\mathcal{G}$-invariant, by the equation

$$
(\mathcal{A}h)_0(z) = 2 \int_{-\infty}^{\infty} 12^{-\frac{i\lambda}{2}} (1 - 2^{i\lambda})(1 - 3^{i\lambda}) \frac{\pi^{\frac{1-i\lambda}{2}}}{\Gamma(\frac{1-i\lambda}{2})} \, \psi(i\lambda) \, E^* \left(z, \frac{1-i\lambda}{2} \right) d\lambda
$$

$$
+ \frac{2}{\sqrt{3}} \, h(0).
\tag{5.5}
$$

Assuming now that h changes to its negative under $\mathcal{G}$, the spectral decomposition of the second one is given by the equation

$$(\mathcal{A}h)_1(z) = 2 \int_{-\infty}^{\infty} 2^{-i\lambda}(1-2^{i\lambda}) \frac{\pi^{\frac{1-i\lambda}{2}}}{\Gamma(\frac{1-i\lambda}{2})} \psi(i\lambda) E^* \left(z, \frac{1-i\lambda}{2}\right) d\lambda - h(0). \quad (5.6)$$

Proof. Choosing $g = \begin{pmatrix} y^{\frac{1}{2}} & x y^{-\frac{1}{2}} \\ 0 & y^{-\frac{1}{2}} \end{pmatrix}$, one obtains

$$(h \circ g^{-1})(q, p) = H\left(\frac{q^2 - 2x\,q\,p + |z|^2\,p^2}{y}\right) = H\left(\frac{|q - p\,z|^2}{y}\right). \quad (5.7)$$

With a view towards applying (4.36), set (noting that, as we must apply (5.1), the two functions of $i\pi\mathcal{E}$ considered below are the transpose of those occurring in (4.36))

$$h_0 = 12^{-i\pi\,\mathcal{E}}\left[1 - 2^{2i\pi\,\mathcal{E}}\right]\left[1 - 3^{2i\pi\,\mathcal{E}}\right]h\,,$$
$$h_1 = 4^{-i\pi\,\mathcal{E}}\left[1 - 2^{2i\pi\mathcal{E}}\right]h\,: \quad (5.8)$$

in other words, one has $h_0(q, p) = H_0(q^2 + p^2)$ and $h_1(q, p) = H_1(q^2 + p^2)$ with

$$H_0(\rho) = \frac{1}{\sqrt{12}}\left[H\left(\frac{\rho}{12}\right) - 2H\left(\frac{\rho}{3}\right) - 3H\left(\frac{3\rho}{4}\right) + 6H(3\rho)\right],$$
$$H_1(\rho) = -H(\rho) + \frac{1}{2}H\left(\frac{\rho}{4}\right). \quad (5.9)$$

We end up with the pair of equations, in which $g.i = z$,

$$(\mathcal{A}h)_0(z) = \sum_{m,n\in\mathbb{Z}} H_0\left(\frac{|m - nz|^2}{y}\right),$$
$$(\mathcal{A}h)_1(z) = \sum_{m,n\in\mathbb{Z}} H_1\left(\frac{|m - nz|^2}{y}\right). \quad (5.10)$$

The functions $(\mathcal{A}h)_0$ and $(\mathcal{A}h)_1$ are clearly Γ-automorphic, and we now make their spectral decompositions explicit.

The function

$$\psi(s) = \frac{1}{2\pi} \int_0^{\infty} r^{-s} H(r^2)\,dr \quad (5.11)$$

is holomorphic for $\mathrm{Re}\ s < 1$: we shall use later the fact that it extends as a meromorphic function in the domain $\mathrm{Re}\ s < 3$ with a simple pole at $s = 1$, and that

$$\psi(-1) = \frac{1}{4\pi^2} \int_{\mathbb{R}^2} h(q, p)\,dq\,dp = \frac{1}{8\pi^2}(\mathcal{G}h)(0). \quad (5.12)$$

One has

$$H(\rho) = \int_{-\infty}^{\infty} \psi(i\lambda)\,\rho^{-\frac{1}{2}+\frac{i\lambda}{2}}\,d\lambda\,, \qquad \rho > 0 : \tag{5.13}$$

an integration by parts based on

$$r^{-s} = \frac{1}{(1-s)\dots(k-s)}\,\frac{d^k}{dr^k}\,(r^{-s+k}) \tag{5.14}$$

shows that the function ψ is rapidly decreasing at infinity on vertical strips with $\mathrm{Re}\ s < 1$, and one can rewrite (5.13) as

$$H(\rho) = \frac{2}{i}\int_{\sigma-i\infty}^{\sigma+i\infty} \psi(1-2s)\,\rho^{-s}\,ds \tag{5.15}$$

for any $\sigma > 0$ (the formula above corresponds to $\sigma = \frac{1}{2}$). Also,

$$\| h \|_{L^2(\mathbb{R}^2)}^2 = 4\pi^2 \int_{-\infty}^{\infty} |\psi(i\lambda)|^2\,d\lambda\,. \tag{5.16}$$

Since the operator $2i\pi\mathcal{E}$ acts on a function homogeneous of degree $-1 + i\lambda$ like the multiplication by $i\lambda$, (5.13) and (5.8) show that

$$h_0(q,\,p) = \int_{-\infty}^{\infty} 12^{-\frac{i\lambda}{2}}\,(1-2^{i\lambda})\,(1-3^{i\lambda})\,\psi(i\,\lambda)\,(q^2+p^2)^{-\frac{1}{2}+\frac{i\lambda}{2}}\,d\lambda\,,$$

$$h_1(q,\,p) = \int_{-\infty}^{\infty} 2^{-i\lambda}\,(1-2^{i\lambda})\,\psi(i\,\lambda)\,(q^2+p^2)^{-\frac{1}{2}+\frac{i\lambda}{2}}\,d\lambda\,. \tag{5.17}$$

From (5.10) and (5.15), it thus follows that, for $\sigma > 1$,

$$(\mathcal{A}h)_0(z) = H_0(0) + \frac{2}{i}\sum_{\substack{m,n\in\mathbb{Z}\\|m|+|n|\neq 0}}\int_{\sigma-i\infty}^{\sigma+i\infty} 12^{s-\frac{1}{2}}\,(1-2^{1-2s})\,(1-3^{1-2s})$$

$$\times\,\psi(1-2s)\left(\frac{|m-n\,z|^2}{y}\right)^{-s}\,ds\,,$$

$$(\mathcal{A}h)_1(z) = H_1(0) + \frac{2}{i}\sum_{\substack{m,n\in\mathbb{Z}\\|m|+|n|\neq 0}}\int_{\sigma-i\infty}^{\sigma+i\infty} -(2^{2s-1}-1) \tag{5.18}$$

$$\times\,\psi(1-2s)\left(\frac{|m-n\,z|^2}{y}\right)^{-s}\,ds\,.$$

Recall that, in the domain $\mathrm{Re}\ s > 1$, one defines the Eisenstein series

$$E(z,\,s) = \frac{1}{2}\sum_{\substack{m,n\in\mathbb{Z}\\(m,n)=1}}\left(\frac{|m-n\,z|^2}{y}\right)^{-s} \tag{5.19}$$

and that, if one sets

$$\zeta^*(s) = \zeta^*(1-s) = \pi^{-\frac{s}{2}} \Gamma(\tfrac{s}{2}) \zeta(s),$$

$$E^*(z, s) = \zeta^*(2s) E(z, s) = \pi^{-s} \Gamma(s) \zeta(2s) E(z, s), \qquad (5.20)$$

the function $s \mapsto E^*(z, s)$ extends as a meromorphic function throughout the complex plane, regular outside the points 0 and 1 which are simple poles, satisfying the functional equation $E^*(z, s) = E^*(z, 1-s)$. As seen from the classical expansion,

$$E^*(z, s) = \zeta^*(2s) y^s + \zeta^*(2s-1) y^{1-s}$$
$$+ 2 \sum_{n \neq 0} |n|^{s-\frac{1}{2}} \sigma_{1-2s}(|n|) K_{s-\frac{1}{2}}(2\pi|n|y) e^{2i\pi nx}, \qquad (5.21)$$

with $\sigma_{1-2s}(|n|) = \sum_{1 \leq d | n} d^{1-2s}$, one has

$$\mathrm{Res}_{s=1} E^*(z, s) = \frac{1}{2}, \qquad \mathrm{Res}_{s=1} E(z, s) = \frac{3}{\pi}. \qquad (5.22)$$

One may then write, using also (5.9),

$$(\mathcal{A}h)_0(z) = 3^{-\frac{1}{2}} h(0)$$
$$+ \frac{4}{i} \int_{\sigma-i\infty}^{\sigma+i\infty} 12^{s-\frac{1}{2}} (1 - 2^{1-2s})(1 - 3^{1-2s}) \zeta(2s) \psi(1-2s) E(z, s) \, ds,$$

$$(\mathcal{A}h)_1(z) = -\frac{1}{2} h(0) + \frac{4}{i} \int_{\sigma-i\infty}^{\sigma+i\infty} (2^{2s-1} - 1) \zeta(2s) \psi(1-2s) E(z, s) \, ds. \qquad (5.23)$$

If one moves the integration line to $\sigma = \frac{1}{2}$, the pole of $\zeta(2s) E(z, s)$ at $s = \frac{1}{2}$ is killed by the factor $1 - 2^{\pm(1-2s)}$. The pole at $s = 1$ originating from the factor $E(z, s)$ contributes to $(\mathcal{A}h)_0(z)$ or $(\mathcal{A}h)_1(z)$ a term easily computed with the help of (5.12) and (5.22), to wit $3^{-\frac{1}{2}} (\mathcal{G}h)(0)$, (resp. $\frac{1}{2} (\mathcal{G}h)(0)$). One thus obtains

$$(\mathcal{A}h)_0(z) = 2 \int_{-\infty}^{\infty} 12^{-\frac{i\lambda}{2}} (1 - 2^{i\lambda})(1 - 3^{i\lambda}) \zeta(1 - i\lambda) \psi(i\lambda) E\left(z, \frac{1-i\lambda}{2}\right) d\lambda$$
$$+ 3^{-\frac{1}{2}} [h(0) + (\mathcal{G}h)(0)] \qquad (5.24)$$

and

$$(\mathcal{A}h)_1(z) = 2 \int_{-\infty}^{\infty} 2^{-i\lambda} (1 - 2^{i\lambda}) \zeta(1 - i\lambda) \psi(i\lambda) E\left(z, \frac{1-i\lambda}{2}\right) d\lambda$$
$$+ \frac{1}{2} [(\mathcal{G}h)(0) - h(0)], \qquad (5.25)$$

two expressions which can be rewritten as (5.5) and (5.6).

The convergence of the integrals that precede is taken care of by the fact that the function zeta is bounded on the set Re $s \geq 1$, $|s - 1| \geq 1$. Another fact, to be used in a moment, is that the function $z \mapsto (s - 1) E(z, \frac{1-i\lambda}{2})$ is, in the usual fundamental domain, a $O(y^{\frac{1}{2}})$, as $y \to \infty$, in a way uniform with respect to λ. Moreover, a classical integration by parts with respect to λ makes it possible to see [12, p. 104] that the function $(\mathcal{A}h)_0$ or $(\mathcal{A}h)_1$ is actually, in the same conditions, a $O(y^{\frac{1}{2}} (\log y)^{-2})$: this shows not only that each of these two functions is square-integrable in the fundamental domain (with respect to the invariant measure $y^{-2} \, dx \, dy$), but makes the integral of such a function against any function $z \mapsto E(z, \frac{1-i\nu}{2})$, $\nu \in \mathbb{R}$, meaningful.

The proof is not quite over. Indeed, since $E^*(z, \frac{1-i\lambda}{2}) = E^*(z, \frac{1+i\lambda}{2})$, the function $\Phi(\lambda)$ against $E^*(z, \frac{1-i\lambda}{2})$ in either integrand on the right-hand side of (5.5) or (5.6) making such an identity true cannot be unique, unless some relation is imposed between $\Phi(\lambda)$ and $\Phi(-\lambda)$: we shall call such an identity a *proper* Roelcke-Selberg decomposition if the function Φ is even. One can then immediately transform the version using an integral on the line (as used for instance in [24]) into the version using an integral on the half-line, as used in [24, p. 254] or [14, Theorem 15.2].

So that the equations (5.5) and (5.6) should qualify as proper Roelcke-Selberg decompositions of automorphic functions, all that remains to be done is substituting for the function against $E^*(z, \frac{1-i\lambda}{2})$ in each integrand its even part. With $\rho = q^2 + p^2$, and $h(q, p) = H(q^2 + p^2)$, (5.13) leads to

$$(\mathcal{F} h)(q, p) = \int_{-\infty}^{\infty} \pi^{-i\lambda} \frac{\Gamma(\frac{1+i\lambda}{2})}{\Gamma(\frac{1-i\lambda}{2})} \rho^{-\frac{1}{2} - \frac{i\lambda}{2}} \, \psi(i\lambda) \, d\lambda, \qquad (5.\dot{2}6)$$

hence

$$(\mathcal{G} h) (q, p) = \int_{-\infty}^{\infty} (2\pi)^{-i\lambda} \frac{\Gamma(\frac{1+i\lambda}{2})}{\Gamma(\frac{1-i\lambda}{2})} \rho^{-\frac{1}{2} - \frac{i\lambda}{2}} \, \psi(i\lambda) \, d\lambda. \qquad (5.27)$$

One may thus write $(\mathcal{G} h)(q, p) = \tilde{H}(q^2 + p^2)$ with

$$\tilde{H}(\rho) = \int_{-\infty}^{\infty} \tilde{\psi}(i\lambda) \rho^{-\frac{1}{2} + \frac{i\lambda}{2}} \, d\lambda, \qquad \rho > 0, \qquad (5.28)$$

where

$$\tilde{\psi}(i\lambda) = (2\pi)^{i\lambda} \frac{\Gamma(\frac{1-i\lambda}{2})}{\Gamma(\frac{1+i\lambda}{2})} \, \psi(-i\lambda). \qquad (5.29)$$

Now, the function $\lambda \mapsto 12^{-\frac{i\lambda}{2}} (1 - 2^{i\lambda}) (1 - 3^{i\lambda})$ is the product of $2^{-\frac{i\lambda}{2}}$ by an even function, and $\lambda \mapsto 2^{-i\lambda} - 1$ is the product of $2^{-\frac{i\lambda}{2}}$ by an odd function. Using (5.29), one sees that the function against $E^*(z, \frac{1-i\lambda}{2})$ in the integrand of (5.5) is already an even function in the case when $\mathcal{G} h = h$, and that the function against $E^*(z, \frac{1-i\lambda}{2})$ in the integrand of (5.6) is already an even function in the case when $\mathcal{G} h = -h$. This means that the right-hand sides of the two equations under discussion will indeed be proper Roelcke-Selberg expansions provided that, in the first one, h is replaced by $\frac{1}{2}(h + \mathcal{G} h)$ and, in the second one, it is replaced by $\frac{1}{2}(h - \mathcal{G} h)$. Now, as has been recalled in Section 2, this will not change the even-even (*resp.* odd-odd) part of $\mathrm{Op}(h)$ in the first (*resp.* second) case. In view of (5.3), the two equations will remain true after this change. $\qquad\square$

Corollary 5.2. *Let $h \in \mathcal{S}(\mathbb{R}^2)$, be a radial function, and assume that $h(0) = (\mathcal{G} h)(0) = 0$. Under the assumptions of Theorem 5.1, and under the additional assumption that $h = \mathcal{G} h$, in the first equation below, and that $h = -\mathcal{G} h$, in the second one, one has the identities*

$$\| (\mathcal{A}h)_0 \|^2_{L^2(\Gamma \backslash \Pi)} = \frac{8}{\pi} \| (1 - 2^{2i\pi\mathcal{E}})(1 - 3^{2i\pi\mathcal{E}}) \zeta(1 - 2i\pi\mathcal{E}) \, h \|^2_{L^2(\mathbb{R}^2)},$$

$$\| (\mathcal{A}h)_1 \|^2_{L^2(\Gamma \backslash \Pi)} = \frac{8}{\pi} \| (1 - 2^{2i\pi\mathcal{E}}) \zeta(1 - 2i\pi\mathcal{E}) \, h \|^2_{L^2(\mathbb{R}^2)}. \tag{5.30}$$

Proof. Under these assumptions, it is also part of the Roelcke-Selberg theorem [14, Theorems 15.2, 15.5] that

$$\int_{\Gamma \backslash \Pi} \overline{E(z, \frac{1 - i\lambda}{2})} \, (\mathcal{A}h)_0(z) \, d\mu(z)$$

$$= 16\pi \times 12^{-\frac{i\lambda}{2}} (1 - 2^{i\lambda})(1 - 3^{i\lambda}) \zeta(1 - i\lambda) \, \psi(i\lambda),$$

$$\int_{\Gamma \backslash \Pi} \overline{E(z, \frac{1 - i\lambda}{2})} \, (\mathcal{A}h)_1(z) \, d\mu(z)$$

$$= 16\pi \times 2^{-i\lambda} (1 - 2^{i\lambda}) \zeta(1 - i\lambda) \, \psi(i\lambda) \tag{5.31}$$

and

$$\| (\mathcal{A}h)_0 \|^2_{L^2(\Gamma \backslash \Pi)} = 32\pi \int_{-\infty}^{\infty} | (1 - 2^{i\lambda})(1 - 3^{i\lambda}) \zeta(1 - i\lambda) \, \psi(i\lambda) |^2 \, d\lambda,$$

$$\| (\mathcal{A}h)_1 \|^2_{L^2(\Gamma \backslash \Pi)} = 32\pi \int_{-\infty}^{\infty} | (1 - 2^{i\lambda}) \zeta(1 - i\lambda) \, \psi(i\lambda) |^2 \, d\lambda. \tag{5.32}$$

Finally, the last pair of equations transforms to (5.30) if one starts from (5.16) and (5.13), observing that the latter identity is a decomposition of $h(q, p) = H(q^2 + p^2)$ as an integral of functions homogeneous of degree $-1 + i\lambda$. $\qquad\square$

Just as an introduction to what may be possible if Γ is replaced by some congruence group (we are not claiming that what follows is an indication of what would happen in the general case), let us start from the Dirac comb $\mathfrak{d}_0$ in place of $\mathfrak{d}_{even}$ or $\mathfrak{d}_{odd}$. This time, we set $\mathfrak{d}_0^{\tilde{g}} = \mathrm{Met}(\tilde{g}^{-1}) \mathfrak{d}_0$ and, if $h \in \mathcal{S}(\mathbb{R}^2)$ satisfies $h(0) = 0$, we consider the expression

$$(\mathfrak{d}_0^{\tilde{g}} \, | \, \mathrm{Op}(h) \, \mathfrak{d}_0^{\tilde{g}}) = \langle h, \, W(\mathfrak{d}_0^{\tilde{g}}, \mathfrak{d}_0^{\tilde{g}}) \rangle$$

$$= \langle h \circ g^{-1}, \, W(\mathfrak{d}_0, \mathfrak{d}_0) \rangle. \tag{5.33}$$

Again, in the case when h is radial, $h(q, p) = H(q^2 + p^2)$, this expression depends only on $z = g.i$ if $\tilde{g}$ lies above $g \in SL(2, \mathbb{R})$, and we denote it as $(\mathcal{B}h)(z)$. We assume that $h = \mathcal{G}h$ and, for simplicity, that $h(0) = 0$. We obtain (using again Theorem 4.1)

$$(\mathcal{B}h)(z) = \frac{1}{2} \sum_{m,n \in \mathbb{Z}} e^{i\pi mn} H\left(\frac{|m - nz|^2}{4y}\right). \tag{5.34}$$

We must this time take the group Γ_2, rather than Γ, into consideration. Note that a matrix $\left(\begin{smallmatrix} a & b \\ c & d \end{smallmatrix}\right) \in SL(2, \mathbb{Z})$ lies in Γ_2 if and only if $\left(\begin{smallmatrix} a & b \\ c & d \end{smallmatrix}\right) \equiv \left(\begin{smallmatrix} 1 & 0 \\ 0 & 1 \end{smallmatrix}\right)$ or $\left(\begin{smallmatrix} a & b \\ c & d \end{smallmatrix}\right) \equiv \left(\begin{smallmatrix} 0 & 1 \\ 1 & 0 \end{smallmatrix}\right) \mod 2$, while $\left(\begin{smallmatrix} a & b \\ c & d \end{smallmatrix}\right)$ lies in Hecke's group denoted as $\Gamma_0(2)$ if and only if $\left(\begin{smallmatrix} a & b \\ c & d \end{smallmatrix}\right) \equiv \left(\begin{smallmatrix} 1 & 0 \\ 0 & 1 \end{smallmatrix}\right)$ or $\left(\begin{smallmatrix} a & b \\ c & d \end{smallmatrix}\right) \equiv \left(\begin{smallmatrix} 1 & 1 \\ 0 & 1 \end{smallmatrix}\right) \mod 2$: hence, the conjugation map $g \mapsto \left(\begin{smallmatrix} 1 & 0 \\ 1 & 1 \end{smallmatrix}\right) g \left(\begin{smallmatrix} 1 & 0 \\ -1 & 1 \end{smallmatrix}\right)$ is an isomorphism from $\Gamma_0(2)$ onto Γ_2.

Since this is a congruence subgroup, we may take advantage of the very complete exposition of the spectral theorem as done in [12]. A fundamental domain D of the group Γ_2 within $SL(2, \mathbb{Z})$) is defined (with $z = x + iy$) by the inequalities $|x| < 1$, $|z| > 1$; there are two inequivalent cusps, to wit $\mathfrak{a} = i\infty$ and $\mathfrak{b} = 1$. We denote as $\Gamma_2(\mathfrak{a})$ (resp. $\Gamma_2(\mathfrak{b})$) the subgroup of Γ_2 stabilizing the corresponding cusp. Hence, $\Gamma_2(\mathfrak{a})$ consists of all matrices $\pm g$, where g lies in the group $\Gamma_2^\circ(\mathfrak{a})$ generated by the matrix $\gamma_\mathfrak{a} = \left(\begin{smallmatrix} 1 & 2 \\ 0 & 1 \end{smallmatrix}\right)$, and $\Gamma_2(\mathfrak{b})$ consists of all matrices $\pm g$, where g lies in the group $\Gamma_2^\circ(\mathfrak{b})$ generated by the matrix $\gamma_\mathfrak{b} = \left(\begin{smallmatrix} 0 & 1 \\ -1 & 2 \end{smallmatrix}\right)$. We then follow [12, p. 42] or [14, p. 355], defining in each case a *scaling matrix* $\sigma_\mathfrak{a}$ or $\sigma_\mathfrak{b}$ in $SL(2, \mathbb{R})$ such that

$$\sigma_\mathfrak{a}^{-1} \gamma_\mathfrak{a} \sigma_\mathfrak{a} = \left(\begin{smallmatrix} 1 & 1 \\ 0 & 1 \end{smallmatrix}\right), \qquad \sigma_\mathfrak{b}^{-1} \gamma_\mathfrak{b} \sigma_\mathfrak{b} = \left(\begin{smallmatrix} 1 & 1 \\ 0 & 1 \end{smallmatrix}\right): \tag{5.35}$$

one finds that $\sigma_\mathfrak{a} = \left(\begin{smallmatrix} 2^{\frac{1}{2}} & 0 \\ 0 & 2^{-\frac{1}{2}} \end{smallmatrix}\right)$ and $\sigma_\mathfrak{b} = \left(\begin{smallmatrix} 1 & -1 \\ 1 & 0 \end{smallmatrix}\right)$ and will do. Next, we note that if $\gamma = \left(\begin{smallmatrix} a & b \\ c & d \end{smallmatrix}\right) \in \Gamma_2$, the class of γ in $\Gamma_2^\circ(\mathfrak{a}) \backslash \Gamma_2$ is characterized by the pair (c, d) subject to the conditions that $(c, d) = 1$ and that cd must be even. Also, the class of γ in $\Gamma_2^\circ(\mathfrak{b}) \backslash \Gamma_2$ is characterized by the pair $(j, k) = (c - a, d - b)$ subject to the conditions that $(q, p) = 1$ and that both j and k must be odd: to see this, observe that $\left(\begin{smallmatrix} 0 & 1 \\ -1 & 2 \end{smallmatrix}\right)^\ell = \left(\begin{smallmatrix} 1-\ell & \ell \\ -\ell & \ell+1 \end{smallmatrix}\right)$ and that the condition $ad - bc = 1$ can be written $ak - dj = 1 - jk$.

One then defines, for Re $s > 1$, [14, p. 388]

$$E_\mathfrak{a}(z, s) = \sum_{\gamma \in \Gamma_\mathfrak{a} \backslash \Gamma} \left(\text{Im}\, [\sigma_\mathfrak{a}^{-1} \gamma](z)\right)^s. \tag{5.36}$$

Explicitly,

$$E_\mathfrak{a}(z, s) = \frac{1}{2} \sum_{\substack{(c,d)=1 \\ cd \equiv 0 \bmod 2}} \left(\frac{\frac{y}{2}}{|cz + d|^2}\right)^s. \tag{5.37}$$

On the other hand, if $\gamma = \left(\begin{smallmatrix} a & b \\ c & d \end{smallmatrix}\right)$, one has $[\sigma_b^{-1}\gamma](z) = \frac{cz+d}{(c-a)z+d-b}$ so that, similarly,

$$E_b(z, s) = \frac{1}{2} \sum_{\substack{(q,p)=1 \\ q, p \, \text{odd}}} \left(\frac{y}{|qz+p|^2} \right)^s . \tag{5.38}$$

We need to make the (symmetric) scattering matrix

$$\Phi = \begin{pmatrix} \phi_{a,a} & \phi_{a,b} \\ \phi_{b,a} & \phi_{b,b} \end{pmatrix} \tag{5.39}$$

fully explicit: this is a special case of computations done in [12], the result of which, in the particular case under study, we now reproduce. Note that

$$\sigma_a^{-1} \left(\begin{smallmatrix} a & b \\ c & d \end{smallmatrix}\right) \sigma_a = \left(\begin{smallmatrix} a & \frac{b}{2} \\ 2c & d \end{smallmatrix}\right),$$

$$\sigma_a^{-1} \left(\begin{smallmatrix} a & b \\ c & d \end{smallmatrix}\right) \sigma_b = \begin{pmatrix} 2^{-\frac{1}{2}}(a+b) & 2^{-\frac{1}{2}}a \\ 2^{\frac{1}{2}}(c+d) & 2^{\frac{1}{2}}d \end{pmatrix},$$

$$\sigma_b^{-1} \left(\begin{smallmatrix} a & b \\ c & d \end{smallmatrix}\right) \sigma_b = \left(\begin{smallmatrix} c+d & -c \\ c-a+d-b & a-c \end{smallmatrix}\right). \tag{5.40}$$

Following (*loc. cit.*, p. 51), one defines in each case a certain set $\mathcal{C}_{a,a}$ (*resp.* $\mathcal{C}_{a,b}$, $\mathcal{C}_{b,b}$) by considering the set of possible lower-left elements ξ of the matrices on the right-hand sides of (5.40) under the conditions that $\xi > 0$ and that the matrix $\left(\begin{smallmatrix} a & b \\ c & d \end{smallmatrix}\right)$ should lie in Γ. One finds that

$$\mathcal{C}_{a,a} = \{2n, \, n \in \mathbb{N}^\times\}, \quad \mathcal{C}_{a,b} = \{2^{\frac{1}{2}}n, \, n \text{ odd} \in \mathbb{N}^\times\}, \quad \mathcal{C}_{b,b} = \{2n, \, n \in \mathbb{N}^\times\}. \tag{5.41}$$

Then, still following (*loc. cit.*, p. 52), one defines the number

$$S_{a,b}(0, 0, \xi) = \# \left\{\eta \bmod \xi: \left(\begin{smallmatrix} * & * \\ \xi & \eta \end{smallmatrix}\right) \in \sigma_a^{-1}\Gamma\sigma_b\right\}, \tag{5.42}$$

and two other sums defined in an analogous way. One finds, denoting as $\phi_{\mathcal{E}}$ the Euler indicator function, that

$$S_{a,a}(0, 0, \xi) = \phi_{\mathcal{E}}(\xi), \quad S_{a,b}(0, 0, \xi) = \phi_{\mathcal{E}}(2^{-\frac{1}{2}}\xi), \quad S_{b,b}(0, 0, \xi) = \phi_{\mathcal{E}}(\xi). \tag{5.43}$$

One then has (*loc. cit.*, p. 66)

$$\phi_{a,b}(s) = \pi^{\frac{1}{2}} \frac{\Gamma(s - \frac{1}{2})}{\Gamma(s)} \sum_{\xi \in \mathcal{C}_{a,b}} \xi^{-2s} S_{a,b}(0, 0, \xi) \tag{5.44}$$

and two related equations. Noting that

$$\phi_{\mathcal{E}}(2n) = \begin{cases} \phi_{\mathcal{E}}(n) & \text{if } n \text{ odd,} \\ 2\,\phi_{\mathcal{E}}(n) & \text{if } n \text{ even,} \end{cases} \tag{5.45}$$

so that

$$\sum_{n\geq1}\frac{\phi_{\mathcal{E}}(2n)}{n^s} = \sum_{n\,\text{odd}\geq1}\frac{\phi_{\mathcal{E}}(n)}{n^s} + \sum_{n\,\text{even}\geq2}\frac{2\,\phi_{\mathcal{E}}(n)}{n^s}$$

$$= \sum_{n\geq1}\frac{\phi_{\mathcal{E}}(n)}{n^s} + \sum_{n\,\text{even}\geq2}\frac{\phi_{\mathcal{E}}(n)}{n^s}$$

$$= \frac{\zeta(s-1)}{\zeta(s)} + \sum_{n\geq1}\frac{\phi_{\mathcal{E}}(2n)}{(2n)^s}$$

$$= (1-2^{-s})^{-1}\frac{\zeta(s-1)}{\zeta(s)} \tag{5.46}$$

and

$$\sum_{n\,\text{odd}\geq1}\frac{\phi_{\mathcal{E}}(n)}{n^s} = \frac{1-2^{1-s}}{1-2^{-s}}\frac{\zeta(s-1)}{\zeta(s)}, \tag{5.47}$$

one obtains

$$\phi_{\mathfrak{a},\mathfrak{a}}(s) = \pi^{\frac12}\frac{\Gamma(s-\frac12)}{\Gamma(s)}\sum_{n\geq1}(2n)^{-2s}\,\phi_{\mathcal{E}}(2n)$$

$$= \pi^{\frac12}\frac{\Gamma(s-\frac12)}{\Gamma(s)}\,2^{-2s}\,(1-2^{-2s})^{-1}\frac{\zeta(2s-1)}{\zeta(2s)}$$

$$= 2^{-2s}\,(1-2^{-2s})^{-1}\frac{\zeta^*(2s-1)}{\zeta^*(2s)}, \tag{5.48}$$

and $\phi_{\mathfrak{b},\mathfrak{b}}(s)$ is the same. Next,

$$\phi_{\mathfrak{a},\mathfrak{b}}(s) = \pi^{\frac12}\frac{\Gamma(s-\frac12)}{\Gamma(s)}\sum_{n\,\text{odd}\geq1}\frac{\phi_{\mathcal{E}}(n)}{(2^{\frac32}n)^{2s}}$$

$$= \pi^{\frac12}\frac{\Gamma(s-\frac12)}{\Gamma(s)}\,2^{-s}\frac{1-2^{1-2s}}{1-2^{-2s}}\frac{\zeta(2s-1)}{\zeta(2s)}$$

$$= 2^{-s}\frac{1-2^{1-2s}}{1-2^{-2s}}\frac{\zeta^*(2s-1)}{\zeta^*(2s)}. \tag{5.49}$$

As a safeguard, one may verify that $\Phi(s)\,\Phi(1-s) = 1$.

Theorem 5.3. *Besides the conditions that $h \in \mathcal{S}(\mathbb{R}^2)$ is radial, $h(q, p) = H(q^2 + p^2)$, and satisfies $h(0) = 0$, assume that $h = \mathcal{G}h$. Then, the Roelcke-Selberg expansion of the Γ_2 -automorphic function $\mathcal{B}h$ (cf. (5.34)) is given by the equation*

$$(\mathcal{B}h)(z) = \int_{-\infty}^{\infty}\zeta(1-i\lambda)\,\psi(i\lambda)\left[2^{\frac{3-3i\lambda}{2}}\,E_{\mathfrak{a}}\left(z,\frac{1-i\lambda}{2}\right)\right.$$

$$\left. + (2-2^{1-i\lambda})\,E_{\mathfrak{b}}\left(z,\frac{1-i\lambda}{2}\right)\right]d\lambda, \tag{5.50}$$

with ψ as defined in (5.11).

Proof. Let us set

$$\beta(s) = \frac{(2\pi)^{1-s}}{\Gamma(1-s)}\,\psi(2s-1)\,, \tag{5.51}$$

so as to have (since $h = \mathcal{G}\,h$, using (5.29))

$$\beta(s) = \beta(1-s)\,. \tag{5.52}$$

Next, let us make the functional equation $E(z,\,s) = \Phi(s)\,E(z,\,1-s)$ (with $E(z,\,s) = \left(\begin{smallmatrix} E_a(z,\,s) \\ E_b(z,\,s) \end{smallmatrix}\right)$) explicit:

$$E_a(z,\,s) = \frac{\zeta^*(2-2s)}{\zeta^*(2s)}\left[\frac{2^{-2s}}{1-2^{-2s}}\,E_a(z,\,1-s) + \frac{2^{-s}\,(1-2^{1-2s})}{1-2^{-2s}}\,E_b(z,\,1-s)\right],$$

$$E_b(z,\,s) = \frac{\zeta^*(2-2s)}{\zeta^*(2s)}\left[\frac{2^{-s}\,(1-2^{1-2s})}{1-2^{-2s}}\,E_a(z,\,1-s) + \frac{2^{-2s}}{1-2^{-2s}}\,E_b(z,\,1-s)\right].$$

$$\tag{5.53}$$

Using (5.34), then (5.15), and assuming $\sigma > 1$ for convergence, we obtain

$$(\mathcal{B}\,h)(z) = \frac{1}{2}\sum_{\substack{m,n\in\mathbb{Z}\\|m|+|n|\neq 0}} e^{i\pi mn}\,H\left(\frac{|m-nz|^2}{4y}\right)$$

$$= \frac{1}{i}\int_{\sigma-i\infty}^{\sigma+i\infty}\psi(1-2s)\,ds\sum_{\substack{m,n\in\mathbb{Z}\\|m|+|n|\neq 0}} e^{i\pi mn}\left(\frac{|m-nz|^2}{4y}\right)^{-s}. \tag{5.54}$$

Set $m = rm_1$, $n = rn_1$ with $r \geq 1$ and $(m_1, n_1) = 1$: since

$$\sum_{r\text{ even}\geq 2} r^{-2s} = 2^{-2s}\,\zeta(2s)\,,\qquad \sum_{r\text{ odd}\geq 1} r^{-2s} = (1-2^{-2s})\,\zeta(2s)\,, \tag{5.55}$$

the sum on the right-hand side can be written as

$$2^{-2s}\,\zeta(2s)\sum_{\substack{m,n\in\mathbb{Z}\\(m,n)=1}}\left(\frac{|m-nz|^2}{4y}\right)^{-s}$$

$$+ (1-2^{-2s})\,\zeta(2s)\sum_{\substack{m,n\in\mathbb{Z}\\(m,n)=1}} e^{i\pi mn}\left(\frac{|m-nz|^2}{4y}\right)^{-s}. \tag{5.56}$$

Also, using (5.37) and (5.38),

$$\sum_{\substack{m,n\in\mathbb{Z}\\(m,n)=1}} \left(\frac{|m-nz|^2}{4y}\right)^{-s} = 2^{2s}\left[2^{s+1}\,E_\mathfrak{a}(z,\,s) + 2\,E_\mathfrak{b}(z,\,s)\right],$$

$$\sum_{\substack{m,n\in\mathbb{Z}\\(m,n)=1}} e^{i\pi mn}\left(\frac{|m-nz|^2}{4y}\right)^{-s} = 2^{2s}\left[2^{s+1}\,E_\mathfrak{a}(z,\,s) - 2\,E_\mathfrak{b}(z,\,s)\right]. \tag{5.57}$$

Hence,

$$\sum_{\substack{m,n\in\mathbb{Z}\\|m|+|n|\neq0}} e^{i\pi mn}\left(\frac{|m-nz|^2}{4y}\right)^{-s} = \zeta(2s)\,\Xi(z,\,s) \tag{5.58}$$

with

$$\Xi(z,\,s) = 2^{s+1}\,E_\mathfrak{a}(z,\,s) + 2\,E_\mathfrak{b}(z,\,s) + (2^{2s}-1)\left[2^{s+1}\,E_\mathfrak{a}(z,\,s) - 2\,E_\mathfrak{b}(z,\,s)\right]$$
$$= 2^{3s+1}\,E_\mathfrak{a}(z,\,s) + (4-2^{2s+1})\,E_\mathfrak{b}(z,\,s). \tag{5.59}$$

Note, making use of (5.51), that

$$\zeta(2s)\,\psi(1-2s) = 2^{-s}\,\zeta^*(2s)\,\beta(1-s). \tag{5.60}$$

Hence,

$$(\mathcal{B}\,h)(z) = \frac{1}{i}\int_{\sigma-i\infty}^{\sigma+i\infty} 2^{-s}\,\zeta^*(2s)\,\beta(1-s)\,\Xi(z,\,s)\,ds. \tag{5.61}$$

From (5.53),

$$\Xi(z,s)$$
$$= \frac{\zeta^*(2-2s)}{\zeta^*(2s)}2^{3s+1}\left[\frac{2^{-2s}}{1-2^{-2s}}E_\mathfrak{a}(z,1-s) + \frac{2^{-s}-2^{1-3s}}{1-2^{-2s}}E_\mathfrak{b}(z,1-s)\right]$$
$$+ \frac{\zeta^*(2-2s)}{\zeta^*(2s)}2^{3s+1}(4-2^{2s+1})\left[\frac{2^{-s}2^{1-3s}}{1-2^{-2s}}E_\mathfrak{a}(z,1-s) + \frac{2^{-2s}}{1-2^{-2s}}E_\mathfrak{b}(z,1-s)\right]$$
$$= \frac{\zeta^*(2-2s)}{\zeta^*(2s)}\left[2^{3-s}E_\mathfrak{a}(z,1-s) + (2^{2s+1}-4)E_\mathfrak{b}(z,1-s)\right]. \tag{5.62}$$

Since $\beta(s) = \beta(1-s)$, the integrand of the integral (5.61) can also be written as

$$2^{-s}\,\zeta^*(2-2s)\,\beta(s)\left[2^{3-s}\,E_\mathfrak{a}(z,\,1-s) + (2^{2s+1}-4)\,E_\mathfrak{b}(z,\,1-s)\right], \tag{5.63}$$

which is the same as $2^{s-1}\,\zeta^*(2-2s)\,\beta(s)\,\Xi(z,\,1-s)$. Thus, the integrand is invariant under the change $s \mapsto 1-s$.

Let us discuss its possible poles in the closed half-plane $\mathrm{Re}\,s \geq \frac{1}{2}$. According to general results [12, Theorem 6.10], the poles of the function $s \mapsto \Xi(z,\,s)$ in the open half-plane are to be looked for among those of the function $\phi_{a,a} = \phi_{b,b}$ there and are simple: $s = 1$ is the only possible one. However, this corresponds to a zero of the factor $\beta(1 - s) = \frac{(2\pi)^s}{\Gamma(s)} \psi(2s - 1)$ since, by assumption,

$$0 = h(0) = (\mathcal{G}\,h)(0) = 2 \int_{\mathbb{R}^2} h(x, \xi)\, dx\, d\xi$$

$$= 4\pi \int_0^\infty H(r^2)\, r\, dr = 8\pi^2\, \psi(-1)\,. \tag{5.64}$$

Next [12, Theorem 6.11], the function $s \mapsto \Xi(z, s)$ has no pole on the line $\mathrm{Re}\, s = \frac{1}{2}$: only, the factor $\zeta^*(2s)$ does of course have a simple one at $s = \frac{1}{2}$. Again, this does not contribute to a singularity in the integrand of (5.61) since, this time, one has $\Xi(z, \frac{1}{2}) = 0$, as it follows from the equation $E(z, s) = \Phi(s)\, E(z, 1 - s)$ together with the fact that $\phi_{a,a}(\frac{1}{2}) = -1$ and $\phi_{a,b}(\frac{1}{2}) = 0$. One can thus move the line of integration to the the value $\sigma = \frac{1}{2}$, finding

$$(\mathcal{B}\,h)(z) = \int_{-\infty}^\infty 2^{-\frac{3}{2}+\frac{i\lambda}{2}}\, \zeta^*(1 - i\,\lambda)\, \beta\left(\frac{1+i\lambda}{2}\right)\, \Xi\left(z, \frac{1-i\lambda}{2}\right)\, d\lambda \tag{5.65}$$

or

$$(\mathcal{B}\,h)(z) = \frac{1}{2} \int_{-\infty}^\infty \zeta(1 - i\lambda)\, \psi(i\lambda)\, \Xi\left(z, \frac{1-i\lambda}{2}\right)\, d\lambda\,. \tag{5.66}$$

Explicitly, this is just the expression announced in (5.50). $\qquad\qquad\square$

Chapter 2

Quantization

In this chapter, we develop the appropriate pseudodifferential calculus suitable for the analysis of operators acting on the space $H_{\tau+1}$, $\tau > -1$, of a representation taken from the prolongation of the projective discrete series of representations of $SL(2, \mathbb{R})$. When $\tau = -\frac{1}{2}$, the representation in question is equivalent, under some intertwining, to the even part of the one-dimensional metaplectic representation, and that obtained when $\tau = \frac{1}{2}$ is equivalent to the odd part of the metaplectic representation.

After having introduced the series of representations, we consider several ways to associate, in a covariant way, operators on $H_{\tau+1}$ to symbols, functions on an appropriate phase space. The reader will probably find the collection of symbolic calculi described in what follows too abundant for his taste: two pairs of symbols (of the same operator) living on the half-plane Π are used (the Berezin-contravariant and covariant symbols on one hand, the active and passive symbols on the other hand), as well as three related species of symbols living on $\mathbb{R}^2$ (the isometric horocyclic, hard and and soft horocyclic symbols). Let us mention at once that only the last three ones are really of importance, and not solely because of Theorem 7.2 below (the isometry property): the reason why, in view of our present arithmetic investigations, one cannot use any species of symbol living on Π will be explained in the beginning of Section 10. On the other hand, the definition of the Berezin calculus is an obvious generalization of the so-called Wick calculus, while the active-passive calculus generalizes Weyl's in a straightforward sense too. On the contrary, the construction of symbols of the horocyclic types is not based on any obvious generalization of an already known calculus. The calculi using Π as a phase space will play only a temporary, but necessary, role, preparing for the construction of the horocyclic calculus: note that the three symbols of a horocyclic type are related in such a simple way that we consider the set as defining just *one* calculus.

Species of symbols are usually produced by pairs. Given a map associating a symbol to an operator, there is a natural map in the reverse direction, to wit the adjoint of the first one, when one considers on one hand some appropriate L^2-space of functions on the phase space, on the other hand the Hilbert space of Hilbert-Schmidt operators on $H_{\tau+1}$. The two species of symbols so related will be called dual to each other in what follows. It is in this way that the Berezin-contravariant and Berezin-covariant symbols are dual species of symbols living on Π, and so are the active and passive symbols. The hard horocyclic and soft horocyclic symbols will be dual of each other in the same way, but they are functions, or distributions, on the plane $\mathbb{R}^2$, not on Π. Exceptionally, a species of symbol qualifies as being its own dual, because the map defining it is an isometry between the two Hilbert spaces, of symbols and operators, just mentioned. This is the case with the even-even and odd-odd parts of the Weyl symbol, and the isometric horocyclic symbol may in this respect be considered as the natural generalization of the Weyl symbol.

"Natural" does not mean that its definition is obvious: it is not, and before defining the horocyclic calculus in Section 7, we need, in Section 6, to recall the connections between the four species of symbols living on Π. One of these formulas – that linking the active and passive symbols of the same operator – will find another, independent, use in Section 11. The isometry property – on top of the covariance, which always holds – is not sufficient to qualify a species of symbol as being the *good* one. One must also be able to perform explicit computations: we have no a priori explanation for the fact that these seem to be possible only when using symbols closely related to some $\mathcal{G}$-invariant species, at least as an intermediate. It is for this reason that the pair of horocyclic hard and soft symbols must be considered as well: the symmetry property satisfied by an isometric horocyclic symbol, while still related to $\mathcal{G}$, is more complicated.

One last point deserves to be mentioned, though it will not concern us in the present work. As there is, to our belief, considerable justification in the assertion that it is the horocyclic calculus, not the calculi with Π as a phase space, that is the "good" calculus of operators $H_{\tau+1}$, the reader may wonder why there does not seem to exist any direct, simple enough, definition of it. Actually, there is such a neat definition: it can be found in [31, definition 9.1] in the case when τ is a half-integer $\geq -\frac{1}{2}$, and could be generalized to all cases when $\tau > -1$. However, it depends not only on the τ-dependent generalization of the metaplectic representation, but also on a partial generalization of the Heisenberg representation: the latter one does not have a τ-dependent version, but its infinitesimal operators Q and P do, and constitute ingredients of a version of the horocyclic calculus in which, just as is the case in the Weyl calculus, it is necessary to consider in place of $H_{\tau+1}$ the direct sum of two such spaces, associated to values of τ differing by 1. Contrary to the isometric horocyclic symbol, which was already introduced in the reference just given, the other two species of horocyclic symbols are new material, and have been found necessary towards the completion of computations in the arithmetic part (in Chapter 3) of the present investigations.

6 Discrete series of $SL(2,\mathbb{R})$ and the hyperbolic half-plane as a phase space

We introduce the two realizations $\mathcal{D}_{\tau+1}$ and $\pi_{\tau+1}$ of representations from the (projective, extended) discrete series of $G = SL(2,\mathbb{R})$. We are quite satisfied with projective representations only, *i.e.*, representations up to phase factors (*cf.* item *(i)* in the list of properties from Proposition 6.1). This is why the group $G = SL(2,\mathbb{R})$, rather than its (universal, in general) cover will suffice for the following description of the discrete series, taken from [31, p. 59–60]: our present normalizations will be the same as the ones from this reference; no claim of originality, of course, is made regarding these matters. One should not feel unduly concerned with the fact that subscripts, or superscripts, take sometimes the value τ, sometimes the value $\tau+1$: this originates from the fact that the parametrizations of the projective discrete series and that of Bessel functions are incompatible. No confusion can arise, though, since in the present work τ does not change values, except in the cases when it takes the special value $\pm\frac{1}{2}$, which has been the subject of special consideration in Chapter 1: indeed, the map $\mathrm{Sq}_{\mathrm{even}}$ (*resp.* $\mathrm{Sq}_{\mathrm{odd}}$) introduced between (3.26) and (3.30) intertwines the representation $\mathcal{D}_{\frac{1}{2}}$ (*resp.* $\mathcal{D}_{\frac{3}{2}}$) with the even (*resp.* odd) part of the metaplectic representation.

Proposition 6.1. *Let τ be a real number > -1, and let $H_{\tau+1}$ be the Hilbert space of all (classes of) measurable functions on the half-line $(0,\infty)$ such that*

$$\|v\|_{\tau+1}^2 := \int_0^\infty |v(s)|^2\, s^{-\tau}\, ds < \infty. \tag{6.1}$$

There exists a unitary projective representation $\pi = \mathcal{D}_{\tau+1}$ of G in $H_{\tau+1}$ with the following properties, in the statement of which $g = \left(\begin{smallmatrix} a & b \\ c & d \end{smallmatrix}\right)$:

(i) *for every pair (g,g_1) of elements of G, the complex number $\pi(gg_1)^{-1}\pi(g)\pi(g_1)$ belongs to the group $\exp(2i\pi\tau\mathbb{Z})$;*
(ii) *for every g with $b < 0$, $\pi(g) = e^{i\pi(\tau+1)}\pi(-g)$;*
(iii) *if $b = 0$, $a > 0$, $(\pi(g)v)(s) = a^{\tau-1}v(a^{-2}s)e^{2i\pi\frac{c}{a}s}$;*
(iv) *if $b > 0$, and $v \in C_0^\infty(]0,\infty[)$,*

$$(\pi(g)v)(s) = e^{-i\pi\frac{\tau+1}{2}}\frac{2\pi}{b}\int_0^\infty v(t)\left(\frac{s}{t}\right)^{\frac{\tau}{2}}\exp\left(2i\pi\frac{ds+at}{b}\right)J_\tau\left(\frac{4\pi}{b}\sqrt{st}\right)dt. \tag{6.2}$$

The scalar product in $H_{\tau+1}$ is $(v|u)_{\tau+1} = \int_0^\infty \bar{v}(s)\,u(s)\,s^{-\tau}\,ds$: we shall dispense with the subscript $\tau+1$ when the meaning is clear.

Proposition 6.2. *Under the assumption that $\tau > 0$, consider the Hilbert space $\widetilde{H}_{\tau+1}$ of all holomorphic functions f in Π with*

$$\|f\|_{\widetilde{H}_{\tau+1}}^2 = \int_\Pi |f(z)|^2(\mathrm{Im}\,z)^{\tau+1}\,d\mu(z) < \infty \tag{6.3}$$

together with the map $v \mapsto f$,

$$f(z) = (4\pi)^{\frac{\tau}{2}} \left(\Gamma(\tau)\right)^{-\frac{1}{2}} z^{-\tau-1} \int_0^\infty v(s)\, e^{-2i\pi s z^{-1}}\, ds\,. \qquad (6.4)$$

The map just defined is an isometry from $H_{\tau+1}$ *onto* $\tilde{H}_{\tau+1}$: *it intertwines the representation* $\mathcal{D}_{\tau+1}$ *of* G *in* $H_{\tau+1}$ *and a representation* $\pi_{\tau+1}$ *of* G *in* $\tilde{H}_{\tau+1}$, *taken from the holomorphic (projective) discrete series, characterized up to scalar factors in the group* $\exp(2i\pi\tau\mathbb{Z})$ *by the fact that*

$$(\pi_{\tau+1}(g)f)(z) = (-cz+a)^{-\tau-1} f\left(\frac{dz-b}{-cz+a}\right) \qquad (6.5)$$

if $c < 0$. *In all this the fractional powers which occur are those associated with the principal determination (that with an imaginary part in* $]0, \pi[$) *of the logarithm in* Π.

Remark 6.1. It would have simplified things slightly if the exponential $e^{-2i\pi s z^{-1}}$ in (6.4), could have been replaced by $e^{2i\pi s z}$: in particular, this would have made the definition of $\psi_z^{\tau+1}$, on the right-hand side of (6.6) below, look simpler. However, this is not possible, as we wish to preserve the *exact* intertwining property stated in Proposition 6.2. This is why, so as to simplify things now and then, we shall sometimes take advantage of the invariance of $d\mu(z)$ under the transformation $z \mapsto -z^{-1}$, and manage so as to have to consider the functions $\psi_{-\frac{1}{z}}^{\tau+1}$ in place of the functions $\psi_z^{\tau+1}$.

The second realization, to wit $\pi_{\tau+1}$, of the representation is probably better-known: however, the Hilbert space $\tilde{H}_{\tau+1}$ on which it is defined is simple to describe only when $\tau > 0$ (or $\tau = 0$, the Hardy space case), which is the case of the (projective) discrete (*i.e.*, square integrable) series proper: for $\tau \leq 0$, this is no longer the case, which is the reason why we have referred to the whole series, in which $\tau > -1$, as a prolongation of the discrete series. The first description works for every such value of τ.

Again, there exists a τ-dependent quadratic transformation from functions on the half-line to functions on the line, reducing to $\mathrm{Sq}_{\mathrm{even}}$ (*resp.* $\mathrm{Sq}_{\mathrm{odd}}$) when $\tau = -\frac{1}{2}$ (*resp.* $\tau = \frac{1}{2}$), useful in certain contexts [31, p. 55]. However, we shall have no use for it here, and we shall consider some symbolic calculi of operators acting on the space $H_{\tau+1}$ itself.

Before we come to the presentation of symbolic calculi of operators on $H_{\tau+1}$ using Π as a phase space, we need to discuss coherent states of the representation $\mathcal{D}_{\tau+1}$. These are the functions

$$\psi_z^{\tau+1}(s) = \frac{(4\pi)^{\frac{\tau+1}{2}}}{(\Gamma(\tau+1))^{\frac{1}{2}}} \left(\mathrm{Im}\left(-\frac{1}{z}\right)\right)^{\frac{\tau+1}{2}} s^\tau\, e^{2i\pi s z^{-1}} : \qquad (6.6)$$

it is easily verified that they are normalized as soon as $\tau > -1$ (a standing assumption). Also, they satisfy the basic property that, given $g \in G$, one has

$$\mathcal{D}_{\tau+1}(g)\, \psi_z^{\tau+1} = \omega\, \psi_{g.z}^{\tau+1} \tag{6.7}$$

for some $\omega \in \mathbb{C}$, depending on (g,z), such that $|\omega| = 1$. We shall prove a more precise version: indeed, we do not care about phase factors depending only on g, but we do care about the dependence on z. As

$$\mathrm{Im}\, \left(-\left(\frac{az+b}{cz+d} \right)^{-1} \right) = \left| a + \frac{b}{z} \right|^{-2} \mathrm{Im}\, \left(-\frac{1}{z} \right), \tag{6.8}$$

it is immediate that (6.7) is a consequence of the following lemma.

Lemma 6.3. *Set* $\phi_z^{\tau+1}(s) = s^\tau\, e^{2i\pi s\bar{z}^{-1}}$ *, so that*

$$\psi_z^{\tau+1} = \frac{(4\pi)^{\frac{\tau+1}{2}}}{(\Gamma(\tau+1))^{\frac{1}{2}}} \left(\mathrm{Im}\, \left(-\frac{1}{z} \right) \right)^{\frac{\tau+1}{2}} \phi_z^{\tau+1}. \tag{6.9}$$

For every $g = \left(\begin{smallmatrix} a & b \\ c & d \end{smallmatrix} \right) \in G$, *there exists a continuous determination of the function* $z \mapsto \left(a + \frac{b}{z} \right)^{-\tau-1}$ *in* Π *and a number* $\omega \in \exp(2i\pi\tau\mathbb{Z})$ *depending only on* g *such that*

$$\mathcal{D}_{\tau+1}(g)\, \phi_z^{\tau+1} = \omega \left(a + \frac{b}{z} \right)^{-\tau-1} \phi_{\frac{az+b}{cz+d}}^{\tau+1} \tag{6.10}$$

for every $z \in \Pi$.

Proof. Dropping the normalization constant which occurred in (6.4) (anyway, it would not be meaningful in the case when $\tau \in]-1, 0]$), we set

$$(\mathcal{L}_\tau v)(z) = z^{-\tau-1} \int_0^\infty v(s)\, e^{-2i\pi s z^{-1}}\, ds. \tag{6.11}$$

The intertwining property of Proposition 6.2 is preserved, and even holds when $\tau \in]-1, 0]$, though of course one does not have an isometry any more, and the image of $\mathcal{L}_\tau$ has no simple characterization when $\tau < 0$ (when $\tau = 0$, one gets the Hardy space rather than a weighted L^2-space of holomorphic functions in Π).

It is immediate that, for every $w \in \Pi$, one has

$$(\mathcal{L}_\tau \phi_w^{\tau+1})(z) = C_\tau\, z^{-\tau-1} \left(\frac{z - \bar{w}}{z\,\bar{w}} \right)^{-\tau-1} \tag{6.12}$$

with

$$C_\tau = \frac{\Gamma(\tau+1)}{(2\pi)^{\tau+1}}\, e^{\frac{i\pi(\tau+1)}{2}}. \tag{6.13}$$

As a first step towards proving (6.10), consider the case when $c < 0$ and $b > 0$. Since the four factors involved in the identity

$$z \left(\frac{z - \bar{w}}{z \, \bar{w}} \right) = (z - \bar{w}) \frac{1}{\bar{w}} \tag{6.14}$$

all lie in Π, one can take it to the $(-\tau - 1)$th power in the obvious sense, getting as a result

$$(\mathcal{L}_\tau \phi_w^{\tau+1})(z) = C_\tau \, (z - \bar{w})^{-\tau-1} \left(\frac{1}{\bar{w}} \right)^{-\tau-1}. \tag{6.15}$$

One has

$$(\mathcal{L}_\tau \mathcal{D}_{\tau+1}(\begin{pmatrix} a & b \\ c & d \end{pmatrix})) \phi_w^{\tau+1})(z) = C_\tau \, (-cz + a)^{-\tau-1} \left(\frac{dz - b}{-cz + a} - \bar{w} \right)^{-\tau-1} (\frac{1}{\bar{w}})^{-\tau-1} : \tag{6.16}$$

since, with $w_1 = \frac{aw+b}{cw+d}$, one has

$$(z - \bar{w}_1) \, (c\bar{w} + d) = (-cz + a) \left(\frac{dz - b}{-cz + a} - \bar{w} \right), \tag{6.17}$$

an identity in which all four factors lie in Π, one has

$$(\mathcal{L}_\tau \mathcal{D}_{\tau+1}(\begin{pmatrix} a & b \\ c & d \end{pmatrix})) \phi_w^{\tau+1})(z) = C_\tau \, (z - \bar{w}_1)^{-\tau-1} \, (c\bar{w} + d)^{-\tau-1} \left(\frac{1}{\bar{w}} \right)^{-\tau-1}. \tag{6.18}$$

In order to prove (6.10), we need to compare this expression to

$$(\mathcal{L}_\tau \phi_{w_1}^{\tau+1})(z) = C_\tau \, (z - \bar{w}_1)^{-\tau-1} \left(\frac{1}{\bar{w}_1} \right)^{-\tau-1}. \tag{6.19}$$

The extra factor is

$$(c\bar{w} + d)^{-\tau-1} (\frac{1}{\bar{w}})^{-\tau-1} \left(\frac{c\bar{w} + d}{a\bar{w} + b} \right)^{\tau+1} : \tag{6.20}$$

again, it is a matter of raising the identity

$$(c\bar{w} + d) \frac{1}{\bar{w}} = \left(a + \frac{b}{\bar{w}} \right) \left(\frac{c\bar{w} + d}{a\bar{w} + b} \right) \tag{6.21}$$

to the power $-\tau - 1$: equation (6.10) is proved the case when $c < 0$ and $b > 0$, provided that the fractional power which occurs there is given the principal determination in Π.

Next, since $\begin{pmatrix} a & b \\ c & d \end{pmatrix} \begin{pmatrix} 0 & 1 \\ -1 & 0 \end{pmatrix}^3 = \begin{pmatrix} b & -a \\ d & -c \end{pmatrix}$, a matrix which describes some neighbourhood of the identity when $b > 0$ and $c < 0$, the matrices for which equation (6.10) has been proved so far generate G: the general case easily follows, but it would be harder to tell exactly which (continuous) determination of the function $z \mapsto (a + \frac{b}{z})^{-\tau-1}$ has to be used in general. $\qquad \square$

The family $(\psi_z^{\tau+1})_{z\in\Pi}$ is a family of coherent states of the space $H_{\tau+1}$ in view of (6.7), together with the fact that it constitutes a total set in the Hilbert space under consideration. In the case when $\tau > 0$, one has a more precise result, to wit the fact [30, p. 179] that

$$\| v \|_{\tau+1}^2 = \frac{\tau}{4\pi} \int_\Pi |(\psi_z^{\tau+1} \, | \, v)|^2 \, d\mu(z), \tag{6.22}$$

where $d\mu(x+iy) = y^{-2} \, dx \, dy$. By polarization, one obtains in this case the resolution of the identity

$$u = \frac{\tau}{4\pi} \int_\Pi (\psi_z^{\tau+1} \, | \, u) \, \psi_z^{\tau+1} \, d\mu(z), \tag{6.23}$$

from which one can recover an operator A in $H_{\tau+1}$ from its matrix elements against the family of coherent states under consideration, by means of the identity

$$(v \, | \, Au) = \left(\frac{\tau}{4\pi}\right)^2 \int_{\Pi\times\Pi} (v \, | \, \psi_w^{\tau+1}) \, (\psi_w^{\tau+1} \, | \, A \, \psi_z^{\tau+1}) \, (\psi_z^{\tau+1} \, | \, u) \, d\mu(w) \, d\mu(z). \tag{6.24}$$

This makes it possible to characterize [29] classes of operators in $H_{\tau+1}$ in terms of properties of their symbols for some appropriate τ-related calculus, for instance the one depending on the active symbol (*cf. infra*). Recovering the whole family of matrix elements from the diagonal ones may, or not, be possible, but one can be sure that, if such an extension exists, it is unique. This is a consequence of (6.6): indeed, if one removes (as in (6.9)) the factor $(\mathrm{Im}\,(-z^{-1}))^{\frac{\tau+1}{2}}$ from this expression of $\psi_z^{\tau+1}(z)$, what remains is an antiholomorphic function of z. It follows that the matrix element $(\psi_w^{\tau+1} \, | \, A \, \psi_z^{\tau+1})$ is the product of $(\mathrm{Im}\,(-w^{-1}) \, \mathrm{Im}\,(-z^{-1}))^{\frac{\tau+1}{2}}$ by a sesquiholomorphic function of the pair (w, z): as such, it is characterized by its values on the diagonal. This explains why, even though the operator – to wit, $\Gamma(\tau + \frac{1}{2} + i\pi\mathcal{E})$ – entering (7.63) is very far from invertible (on vertical lines, the Gamma function decreases exponentially at infinity), the map $\mathcal{C}$ which enters that equation must still be one-to-one. At least in the case when $\tau = \pm\frac{1}{2}$, this last property continues to hold even if one takes the space of all even $\mathcal{G}$-invariant tempered distributions as the domain of $\mathcal{C}$.

The consideration of the diagonal matrix elements leads to Berezin's quantization theory, the most popular quantization rule on hermitian symmetric spaces. In the case under study, our modification from Berezin's original definition [3] stems from our use of the space $H_{\tau+1}$ rather than $\widetilde{H}_{\tau+1}$. Given an operator $A: H_{\tau+1} \to H_{\tau+1}$, the Berezin-covariant symbol (a possibly unfortunate terminology: this adjective has nothing to do with the covariance property below, which all symbols to be introduced do enjoy) of A is the function f^{cov} such that

$$f^{\mathrm{cov}}(z) = (\psi_z^{\tau+1} \, | \, A \, \psi_z^{\tau+1}), \qquad z \in \Pi. \tag{6.25}$$

It is immediate, because of (6.7), that the map $A \mapsto f^{\text{cov}}$ just defined satisfies the following covariance property: given $g \in G$, the Berezin-covariant symbol of the operator $\mathcal{D}_{\tau+1}(g)\, A\, (\mathcal{D}_{\tau+1}(g))^{-1}$ is the function $f^{\text{cov}} \circ g^{-1}$, where $(f \circ g^{-1})(z) = f(\frac{dz-b}{-cz+a})$ if $g = \left(\begin{smallmatrix} a & b \\ c & d \end{smallmatrix}\right)$.

Obviously, very little has to be demanded from an operator A so that its Berezin-covariant symbol might be meaningful: it is enough that A should act from the space linearly generated by the functions $\psi_z^{\tau+1}$ to the algebraic dual of that space. This is both an advantage and an inconvenience: for it implies that the map $A \mapsto f^{\text{cov}}$ is very far from being invertible as a map from the space of Hilbert-Schmidt operators in $H_{\tau+1}$ to $L^2(\Pi)$. The Berezin-covariant symbol must be coupled with a dual species, defined as follows: an operator A in $H_{\tau+1}$ is said to admit a Berezin-contravariant symbol f^{contra} if it can be defined by the equation

$$Av = \int_{\Pi} f^{\text{contra}}(z)\, (\psi_z^{\tau+1} \,|\, v)\, \psi_z^{\tau+1}\, d\mu(z) \tag{6.26}$$

representing Av as an integral superposition of the coherent states. This time, very few operators can have a Berezin-contravariant symbol since an operator which admits one is, in many senses, too good! The Berezin-contravariant and Berezin-covariant symbols are a pair of dual symbols in the sense given in the introduction of the present chapter. There is a map Λ such that, given an operator A admitting a contravariant symbol, its covariant symbol is linked to the former one by the equation $f^{\text{cov}} = \Lambda\, f^{\text{contra}}$. There is no need to make it explicit immediately, since it will appear in a moment as the product of three explicit operators commuting with one another. The operator linking the two species of Berezin-type symbols is known in the higher-rank analogous situation [3], [33], but we shall have no need for it here.

There is another pair of dual symbols, still leaving on Π, defined with the help of the Hankel transformation σ^τ such that

$$(\sigma^\tau v)(s) = 2\pi \int_0^\infty v(t) \left(\frac{s}{t}\right)^{\frac{\tau}{2}} J_\tau \left(\frac{4\pi}{b} \sqrt{st}\right) dt. \tag{6.27}$$

Note that it agrees, up to some phase factor, with the transformation

$$\mathcal{D}_{\tau+1}\left(\left(\begin{smallmatrix} 0 & 1 \\ -1 & 0 \end{smallmatrix}\right)\right)$$

as defined in (6.2). Since it commutes with all transformations $\mathcal{D}_{\tau+1}(k)$ with $k \in SO(2)$, the operator $\mathcal{D}_{\tau+1}(g)\, \sigma^\tau\, \mathcal{D}_{\tau+1}(g)^{-1}$, with $g \in G$, only depends on $z = g.i$ and can thus be denoted as σ_z^τ. The operator A in $H_{\tau+1}$ with active symbol f^{act} is then defined as the weakly convergent integral

$$A = 2 \int_{\Pi} f^{\text{act}}\, \sigma_z^\tau\, d\mu(z) : \tag{6.28}$$

in the reverse direction, one defines the passive symbol f^{pass} of an operator A, say of trace class, by the equation

$$f^{\text{pass}}(z) = 2\,\text{Tr}\,(A\,\sigma_z^\tau). \tag{6.29}$$

Again, the active and passive symbols constitute a pair of dual species of symbols according to what has been explained in the introduction of the present chapter. The corresponding calculus has been studied in [26, 29]. Contrary to the Berezin calculus, it may be considered as a pseudodifferential calculus of operators on $H_{\tau+1}$: the way properties of operators correspond to reasonable properties of symbols can be analyzed in a way recalling pseudodifferential analysis, making in particular the composition of symbols a well-defined operation. However, the two calculi are just as bad when arithmetic is concerned, in which case we shall have to turn to the horocyclic calculus described in next section. Meanwhile, let us recall the links between the various calculi introduced so far, since some use will be made of all the formulas to follow.

The operator linking one species of symbol, on Π, to another species of symbol of the same operator, still living on Π, always commutes with the quasi-regular action of G in $L^2(\Pi)$ (this is the one such that, under $g = \left(\begin{smallmatrix} a & b \\ c & d \end{smallmatrix} \right) \in G$, a function f on Π transforms to $f \circ g^{-1}$, with $(f \circ g^{-1})(z) = f\left(\frac{dz-b}{-cz+a} \right)$), as it follows from the fact that all maps from symbols to operators or from operators to symbols introduced in this context are covariant. As such, it must be, in the spectral-theoretic sense, a function of the hyperbolic Laplacian Δ: all formulas below make sense since, in $L^2(\Pi)$, Δ has a purely continuous spectrum coinciding with the interval $[\frac{1}{4}, \infty[$.

Let us start with the operator F_τ giving the passive symbol of an operator in terms of its active symbol [26]. It has a simple integral kernel, to wit the function $(z, w) \mapsto \frac{2 \exp(-\tau d(z,w))}{\sinh d(z,w)}$ (involving the hyperbolic distance d), and provides a resolvent of the Laplacian since (*loc. cit.*) one has

$$(4\pi)^{-2} \left[\Delta + \tau(\tau + 1) \right] F_\tau \, F_{\tau+1} = I. \tag{6.30}$$

Explicitly, it is given by the formula (*loc. cit.*, or [30, (17.15)])

$$F_\tau = 2\pi \frac{\Gamma\left(\frac{\tau}{2} + \frac{1}{4} + \frac{i}{2}\sqrt{\Delta - \frac{1}{4}} \right) \Gamma\left(\frac{\tau}{2} + \frac{1}{4} - \frac{i}{2}\sqrt{\Delta - \frac{1}{4}} \right)}{\Gamma\left(\frac{\tau}{2} + \frac{3}{4} + \frac{i}{2}\sqrt{\Delta - \frac{1}{4}} \right) \Gamma\left(\frac{\tau}{2} + \frac{3}{4} - \frac{i}{2}\sqrt{\Delta - \frac{1}{4}} \right)}. \tag{6.31}$$

To complete our set of formulas, it suffices to note that the operator expressing the active symbol in terms of the Berezin-contravariant symbol (in the case when such a symbol exists) is self-adjoint in $L^2(\Pi)$, is the same as the operator from the passive symbol to the Berezin-covariant symbol and is given explicitly [30, p. 181], as

$$\frac{2^\tau \pi^{-\frac{1}{2}}}{\Gamma(\tau+1)} \Gamma\left(\frac{\tau}{2} + \frac{3}{4} + \frac{i}{2}\sqrt{\Delta - \frac{1}{4}} \right) \Gamma\left(\frac{\tau}{2} + \frac{3}{4} - \frac{i}{2}\sqrt{\Delta - \frac{1}{4}} \right). \tag{6.32}$$

Combining the results which precede, one obtains the link from the active symbol to the covariant symbol: it is expressed by means of the operator

$$\frac{2^{\tau+1}\,\pi^{\frac{1}{2}}}{\Gamma(\tau+1)}\,\Gamma\left(\frac{\tau}{2}+\frac{1}{4}+\frac{i}{2}\sqrt{\Delta-\frac{1}{4}}\right)\,\Gamma\left(\frac{\tau}{2}+\frac{1}{4}-\frac{i}{2}\sqrt{\Delta-\frac{1}{4}}\right)\,. \tag{6.33}$$

To put some order in the preceding set of symbols and equations, let us note the following. Going from the most "irregular" type of symbol to the most regular one, one finds in succession the contravariant, active, passive and covariant types. The operator from the active to the passive symbol is bounded in $L^2(\Pi)$ as soon as $\tau > 0$, and defined for all values of $\tau > -1$ (a standing assumption): its inverse, though not bounded, is the product of $\Delta + \tau(\tau+1)$ by a bounded operator. The gap between the contravariant and active types of symbols, or that between the passive and covariant types, a fortiori that between the contravariant and covariant types, is huge from the point of view of analysis because of the behaviour of the Gamma function at infinity on vertical lines in the complex plane.

When we come to arithmetic, our problems will occur at finite points on the spectrum of Δ, not at infinity, and all these types will have to be replaced by a completely different one, the horocyclic symbol, living on $\mathbb{R}^2$ rather than Π. This will concern us in the next section.

7   Reinstalling $\mathbb{R}^2$ as a phase space: the horocyclic calculus

It is impossible to obtain a manageable calculus with the property that the map from symbols to operators should be an isometry from $L^2(\Pi)$ to the space of Hilbert-Schmidt operators on $H_{\tau+1}$. To solve the problem – which will at the same time have other considerable advantages – one has to replace the half-plane by $\mathbb{R}^2$, taking advantage of the Radon transformation V from functions f on Π to even functions on $\mathbb{R}^2\backslash\{0\}$, defined by the equation [30, p. 25]

$$(Vf)(x,\,\xi) = \int_{-\infty}^{\infty} f\left(\left(\left(\begin{smallmatrix} x & b \\ \xi & d \end{smallmatrix}\right)\left(\begin{smallmatrix} 1 & t \\ 0 & 1 \end{smallmatrix}\right)\right)\cdot i\right)\,dt\,, \qquad \left(\begin{smallmatrix} x & b \\ \xi & d \end{smallmatrix}\right) \in SL(2,\mathbb{R})\,. \tag{7.1}$$

To see that this makes sense provided convergence is ensured, note that if both $\left(\begin{smallmatrix} x & b \\ \xi & d \end{smallmatrix}\right)$ and $\left(\begin{smallmatrix} x & b' \\ \xi & d' \end{smallmatrix}\right)$ lie in G, so that $b = b' + xs$, $d = d' + \xi s$ for some $s \in \mathbb{R}$, one has $\left(\begin{smallmatrix} x & b \\ \xi & d \end{smallmatrix}\right)\left(\begin{smallmatrix} 1 & t \\ 0 & 1 \end{smallmatrix}\right) = \left(\begin{smallmatrix} x & b' \\ \xi & d' \end{smallmatrix}\right)\left(\begin{smallmatrix} 1 & t+s \\ 0 & 1 \end{smallmatrix}\right)$. The Radon transformation has been studied in a very general context by Helgason [11]. The above-given reference details the case of $SL(2,\mathbb{R})$, to be used here.

Let us emphasize again that using $\mathbb{R}^2$, rather than Π, as a phase space, will prove essential in Sections 10 and 11, for reasons not related to the isometry property only. Recall that the operator $2i\pi\mathcal{E}$ has been defined in (2.6), and consider on functions on $\mathbb{R}^2$ the operator

$$T = \left(\frac{\pi}{2}\right)^{\frac{1}{2}} \frac{\Gamma(\frac{1}{2} - i\pi\mathcal{E})}{\Gamma(-i\pi\mathcal{E})} = \pi^{-\frac{1}{2}} \left(-i\pi\mathcal{E}\right) \int_0^\infty t^{-\frac{1}{2}} (1+t)^{-1+i\pi\mathcal{E}} \, dt : \qquad (7.2)$$

the second factor under the integral sign has been defined in (2.7).

Definition 7.1. Given an operator in $H_{\tau+1}$ with an active symbol f^{act}, one defines the isometric horocyclic symbol of this operator as the even function h^{iso} such that

$$h^{\mathrm{iso}} = \mathcal{R}_{\tau+1} \, TV \, f^{\mathrm{act}}, \qquad (7.3)$$

where

$$\mathcal{R}_{\tau+1} = (2\pi)^{\frac{1}{2}} \, \pi^{i\pi\mathcal{E}} \frac{\Gamma(\frac{\tau}{2} + \frac{1}{4} - \frac{i\pi}{2}\mathcal{E})}{\Gamma(\frac{\tau}{2} + \frac{3}{4} + \frac{i\pi}{2}\mathcal{E})}. \qquad (7.4)$$

Of course, the isometric horocyclic symbol still enjoys a covariance property: this time, however, G acts on $\mathbb{R}^2$ (instead of Π) by linear (instead of fractional-liner) transformations. Its fundamental advantage is expressed in the following [30, p. 182]:

Theorem 7.2. *For every $\tau > -1$, the map $A \mapsto h^{\mathrm{iso}}$ from an operator in $H_{\tau+1}$ to its isometric horocyclic symbol is an isometry from the space of Hilbert-Schmidt operators on $H_{\tau+1}$ onto the subspace of $L^2_{\mathrm{even}}(\mathbb{R}^2)$ consisting of all functions invariant under the (unitary) symmetry*

$$\frac{\Gamma(i\pi\mathcal{E}) \, \Gamma(\tau + \frac{1}{2} - i\pi\mathcal{E})}{\Gamma(-i\pi\mathcal{E}) \, \Gamma(\tau + \frac{1}{2} + i\pi\mathcal{E})} \, \mathcal{G}. \qquad (7.5)$$

Also, in the case when $\tau = -\frac{1}{2}$ (*resp.* $\tau = \frac{1}{2}$), and A is an operator in $H_{\tau+1}$, let A_1 be the even-even operator coinciding with $\mathrm{Sq}_{\mathrm{even}} \, A \, \mathrm{Sq}_{\mathrm{even}}^{-1}$ on $L^2_{\mathrm{even}}(\mathbb{R})$ (*resp.* the odd-odd operator coinciding with $\mathrm{Sq}_{\mathrm{odd}} \, A \, \mathrm{Sq}_{\mathrm{odd}}^{-1}$ on $L^2_{\mathrm{odd}}(\mathbb{R})$). Then, the isometric horocyclic symbol of A is just the Weyl symbol of A_1. This will be proved as part of Proposition 11.1.

Taking advantage of (6.33), we may obtain an isometric horocyclic symbol directly from the corresponding Berezin-covariant symbol, through the equation

$$h^{\mathrm{iso}} = (2\pi)^{\frac{1}{2}} \, \pi^{i\pi\mathcal{E}} \frac{\Gamma(\frac{\tau}{2} + \frac{1}{4} - \frac{i\pi}{2}\mathcal{E})}{\Gamma(\frac{\tau}{2} + \frac{3}{4} + \frac{i\pi}{2}\mathcal{E})}, \qquad (7.6)$$

$$TV \left[\left(\frac{2^{\tau+1} \pi^{\frac{1}{2}}}{\Gamma(\tau+1)} \Gamma\left(\frac{\tau}{2} + \frac{1}{4} + \frac{i}{2}\sqrt{\Delta - \frac{1}{4}}\right) \Gamma\left(\frac{\tau}{2} + \frac{1}{4} - \frac{i}{2}\sqrt{\Delta - \frac{1}{4}}\right) \right)^{-1} f^{\mathrm{cov}} \right].$$

Using the equation [30, p. 27] $TV\left(\Delta - \frac{1}{4}\right) = \pi^2\,\mathcal{E}^2\,TV$, a nice feature of the Radon transform which will be used time and again, one can transform this equation into

$$
h^{\text{iso}} = 2^{-\tau-\frac{1}{2}}\,\Gamma(\tau+1)\,\pi^{i\pi\mathcal{E}}\,\left(\Gamma\left(\frac{\tau}{2} + \frac{1}{4} + \frac{i\pi\mathcal{E}}{2}\right)\Gamma\left(\frac{\tau}{2} + \frac{3}{4} + \frac{i\pi\mathcal{E}}{2}\right)\right)^{-1}\,TV\,f^{\text{cov}}
$$

$$
= 2^{-1+i\pi\mathcal{E}}\,\pi^{-\frac{1}{2}+i\pi\mathcal{E}}\,\frac{\Gamma(\tau+1)}{\Gamma(\tau + \frac{1}{2} + i\pi\mathcal{E})}\,TV\,f^{\text{cov}}\,. \tag{7.7}
$$

We must now learn how to compute horocyclic symbols of given operators. In view of the all-important role of the operator $\mathcal{E}$, one starts from the decomposition of even functions on $\mathbb{R}^2$ into their homogeneous components:

$$
h = \int_{-\infty}^{\infty} h_\lambda\,d\lambda\,, \tag{7.8}
$$

with

$$
h_\lambda(x,\,\xi) = \frac{1}{2\pi}\int_0^{\infty} t^{i\lambda}\,h(tx,\,t\xi)\,dt\,. \tag{7.9}
$$

The function h_λ is even and homogeneous of degree $-1 - i\lambda$: hence, it can be recovered from the function $h_\lambda^\flat$ of one variable only, defined as

$$
h_\lambda^\flat(s) = h_\lambda(s,\,1)\,, \tag{7.10}
$$

by the equation

$$
h_\lambda(x,\,\xi) = |\xi|^{-1-i\lambda}\,h_\lambda^\flat\left(\frac{x}{\xi}\right)\,. \tag{7.11}
$$

The following equation [30, p. 29] makes it possible to obtain the homogeneous components of the TV-transform of a function f on Π:

$$
(TV\,f)_\lambda^\flat(s) = \frac{1}{2}\,(2\pi)^{-\frac{3}{2}}\,\frac{\Gamma(\frac{1+i\lambda}{2})}{\Gamma(\frac{i\lambda}{2})}\int_\Pi\left(\frac{|z-s|^2}{y}\right)^{-\frac{1}{2}-\frac{i\lambda}{2}}\,f(z)\,d\mu(z)\,. \tag{7.12}
$$

We are through with the necessary quotations, and proceed further. It is useful to extend the space $H_{\tau+1}$ and the representation $\mathcal{D}_{\tau+1}$ to a distribution setting. Since

$$
(v\,|\,\mathcal{D}_{\tau+1}(g)\,u) = ((\mathcal{D}_{\tau+1}(g))^{-1}\,v\,|\,u) \tag{7.13}
$$

for any pair $(u,\,v)$ in $H_{\tau+1}$ and any $g \in G$, we make take this as a definition of $\mathcal{D}_{\tau+1}(g)\,u$ when $u \in C_{\tau+1}^{-\infty}(\mathbb{R}^+)$, the space of distributions dual of the space $C_{\tau+1}^\infty(\mathbb{R}^+)$ of C^∞-vectors of the representation $\mathcal{D}_{\tau+1}$: indeed, this latter space is invariant under this representation. The operator (s) of multiplication by s and the operator $s\frac{d}{ds} + \frac{1-\tau}{2}$ lie among the infinitesimal operators of the representation under study, since

$$(s) = \frac{1}{2i\pi} \frac{d}{dt}\Big|_{t=0} \mathcal{D}_{\tau+1}\left(\left(\begin{smallmatrix} 1 & 0 \\ t & 1 \end{smallmatrix}\right)\right),$$

$$s\frac{d}{ds} + \frac{1-\tau}{2} = \frac{d}{dt}\Big|_{t=0} \mathcal{D}_{\tau+1}\left(\left(\begin{smallmatrix} e^{-\frac{t}{2}} & 0 \\ 0 & e^{\frac{t}{2}} \end{smallmatrix}\right)\right). \tag{7.14}$$

It follows that the space of distributions $C_{\tau+1}^{-\infty}(\mathbb{R}^+)$ contains every Dirac mass at some point in $]0, \infty[$ as well as every series of such masses, taken on some arithmetic progression, with coefficients making up a slowly increasing family: this is the typical situation that will be encountered.

When dealing with distributions, one has to be a little bit careful about the role of the measure $dm(s) = s^{-\tau} ds$. A function u on $(0, \infty)$ gives rise to the linear form

$$v \mapsto \langle u, v \rangle_{\tau+1} = \int_0^\infty v(s)\, u(s)\, s^{-\tau}\, ds. \tag{7.15}$$

This extends to the case when u is replaced by a measure M on the real line supported in $]0, \infty[$, just setting

$$\langle M, v \rangle_{\tau+1} = \langle M, s \mapsto s^{-\tau} v(s) \rangle. \tag{7.16}$$

In this way, the measure $M = \delta_a$ on $\mathbb{R}^+$, with $a > 0$, gives rise to the linear form

$$v \mapsto \langle \delta_a \,|\, v \rangle_{\tau+1}$$
$$= (s^{-\tau} v(s))(s = a) = a^{-\tau} v(a). \tag{7.17}$$

Given $a, b > 0$, we then consider the operator $Q_{a,b}^\tau$,

$$Q_{a,b}^\tau w = \langle \delta_a, w \rangle_{\tau+1} \delta_b, \tag{7.18}$$

weakly defined by the equation

$$(v \,|\, Q_{a,b}^\tau w)_{\tau+1} = (ab)^{-\tau}\, \bar{v}(b)\, w(a). \tag{7.19}$$

For the time being, the computations involving punctual masses will be used essentially as a way to avoid carrying extra integrals not really concerned with the heavy calculations that follow: later on, in connection with arithmetic, we shall have to use discrete measures in an essential way.

Given $u, v, w \in H_{\tau+1}$, set $P_{v,u} w = (v \,|\, w)_{\tau+1}\, u$, with $(v \,|\, w)_{\tau+1} = \int_0^\infty \bar{v}(a) \langle \delta_a, w \rangle_{\tau+1}\, da$. Then,

$$P_{v,u} w = \int_0^\infty \int_0^\infty \bar{v}(a)\, u(b)\, Q_{a,b}^\tau w\, da\, db. \tag{7.20}$$

The covariant symbol of $P_{v,\,u}$ is the function

$$
\begin{aligned}
f(z) &= (\psi_z^{\tau+1} \mid P_{v,\,u}\,\psi_z^{\tau+1})_{\tau+1} \\
&= (\psi_z^{\tau+1} \mid (v \mid \psi_z^{\tau+1})_{\tau+1}\,u)_{\tau+1} \\
&= (v \mid \psi_z^{\tau+1})_{\tau+1}\,(\psi_z^{\tau+1} \mid u)_{\tau+1} \\
&= \int_0^\infty \int_0^\infty \bar{v}(a)\,u(b)\,(ab)^{-\tau}\,\psi_z^{\tau+1}(a)\,\overline{\psi_z^{\tau+1}}(b)\,da\,db\,.
\end{aligned}
\tag{7.21}
$$

On the other hand, from (7.20), the covariant symbol $z \mapsto (\psi_z^{\tau+1} \mid Q_{a,\,b}^\tau\,\psi_z^{\tau+1})_{\tau+1}$ of $Q_{a,\,b}^\tau$ ought to be defined so that the covariant symbol of $P_{v,\,u}$ should appear as the integral of this covariant symbol against $\bar{v}(a)\,u(b)\,da\,db$. Using (7.21), we must finally define the covariant symbol of $Q_{a,\,b}^\tau$ as

$$
f(z) = (ab)^{-\tau}\,\overline{\psi_z^{\tau+1}}(b)\,\psi_z^{\tau+1}(a)\,,
\tag{7.22}
$$

in other words

$$
f(z) = \frac{(4\pi)^{\tau+1}}{\Gamma(\tau+1)}\left(\operatorname{Im}\left(-\frac{1}{z}\right)\right)^{\tau+1}\exp\left(2i\pi\left(\frac{a}{\bar{z}}-\frac{b}{z}\right)\right)\,.
\tag{7.23}
$$

We now wish to find the homogeneous components of the isometric horocyclic symbol h^{iso} of $Q_{a,\,b}^\tau$, a task for the completion of which we shall have to assume that $\tau > -\frac{1}{2}$. As a special case of (6.7), one has

$$
\mathcal{D}_{\tau+1}\left(\left(\begin{smallmatrix} 0 & 1 \\ -1 & 0 \end{smallmatrix}\right)\right)\psi_z^{\tau+1} = \omega\,\psi_{-\frac{1}{z}}^{\tau+1}
\tag{7.24}
$$

for some ω with $|\omega| = 1$. It follows that, if one sets

$$
\tilde{Q}_{a,\,b}^\tau = \mathcal{D}_{\tau+1}\left(\left(\begin{smallmatrix} 0 & 1 \\ -1 & 0 \end{smallmatrix}\right)\right)Q_{a,\,b}^\tau\left(\mathcal{D}_{\tau+1}\left(\left(\begin{smallmatrix} 0 & 1 \\ -1 & 0 \end{smallmatrix}\right)\right)\right)^{-1}\,,
\tag{7.25}
$$

the covariant symbol of $\tilde{Q}_{a,\,b}^\tau$ is the function

$$
\tilde{f}(z) = \frac{(4\pi)^{\tau+1}}{\Gamma(\tau+1)}\,(\operatorname{Im} z)^{\tau+1}\,e^{2i\pi\,(bz-a\bar{z})}\,,
\tag{7.26}
$$

at least a typographically simpler function. From the covariance of the horocyclic calculus, the isometric horocyclic symbols h^{iso} and $\tilde{h}^{\mathrm{iso}}$ of $Q_{a,\,b}^\tau$ and $\tilde{Q}_{a,\,b}^\tau$ are linked by the relation

$$
\tilde{h}^{\mathrm{iso}}(x,\,\xi) = h^{\mathrm{iso}}(-\xi,\,x)\,.
\tag{7.27}
$$

Since

$$
h^{\mathrm{iso}}(x,\,\xi) = \int_{-\infty}^\infty (h^{\mathrm{iso}})_\lambda(x,\,\xi)\,d\lambda = \int_{-\infty}^\infty |\xi|^{-1-i\lambda}\,(h^{\mathrm{iso}})_\lambda^{\flat}\left(\frac{x}{\xi}\right)d\lambda\,,
\tag{7.28}
$$

it is immediate, from (7.27), that

$$(\tilde{h}^{\mathrm{iso}})^{\flat}_{\lambda}(s) = |s|^{-1-i\lambda}\,(h^{\mathrm{iso}})^{\flat}_{\lambda}(-s^{-1}).\tag{7.29}$$

We first assume $a \neq b$: the case when $a = b$ will be treated as a limiting case, in the proof of Theorem 7.4 below. Recalling (7.12), and using (7.26), one obtains

$$(TV\,\tilde{f})^{\flat}_{\lambda}(s) = 2^{2\tau-\frac{1}{2}}\,\pi^{\tau-\frac{1}{2}}\,\frac{\Gamma(\frac{1+i\lambda}{2})}{\Gamma(\tau+1)\,\Gamma(\frac{i\lambda}{2})}$$

$$\int_0^\infty y^{\tau-\frac{1}{2}+\frac{i\lambda}{2}}\,dy \int_{-\infty}^\infty [(x-s)^2 + y^2]^{-\frac{1}{2}-\frac{i\lambda}{2}}\,e^{2i\pi\,[(b-a)x + i\,(b+a)y]}\,dx \;:\quad (7.30)$$

performing the change of variable $x \mapsto x + s$ and using [16, p. 401]

$$\int_{-\infty}^\infty (x^2+y^2)^{-\frac{1}{2}-\frac{i\lambda}{2}}\,e^{2i\pi(b-a)x}\,dx = \frac{2\,\pi^{\frac{1+i\lambda}{2}}}{\Gamma(\frac{1+i\lambda}{2})}\,|a-b|^{\frac{i\lambda}{2}}\,y^{-\frac{i\lambda}{2}}\,K_{\frac{i\lambda}{2}}(2\pi\,|a-b|\,y),$$

$$\tag{7.31}$$

we find, if $\tau > -\frac{1}{2}$,

$$(TV\,\tilde{f})^{\flat}_{\lambda}(s) = \frac{2^{2\tau+\frac{1}{2}}\,\pi^{\tau+\frac{i\lambda}{2}}}{\Gamma(\tau+1)\,\Gamma(\frac{i\lambda}{2})}\,|a-b|^{\frac{i\lambda}{2}}\,e^{2i\pi(b-a)s}$$

$$\int_0^\infty y^{\tau-\frac{1}{2}}\,e^{-2\pi(a+b)y}\,K_{\frac{i\lambda}{2}}(2\pi\,|a-b|\,y)\,dy.\tag{7.32}$$

According to [16, p. 92], the last integral is

$$2^{-1-2\tau}\,\pi^{-\tau}\Gamma\left(\tau + \frac{1-i\lambda}{2}\right)\Gamma\left(\tau + \frac{1+i\lambda}{2}\right)(ab)^{-\frac{\tau}{2}}\,|a-b|^{-\frac{1}{2}}\mathfrak{P}^{-\tau}_{-\frac{1}{2}+\frac{i\lambda}{2}}\left(\frac{a+b}{|a-b|}\right),$$

$$\tag{7.33}$$

so that, in that case,

$$(TV\,\tilde{f})^{\flat}_{\lambda}(s) = 2^{-\frac{1}{2}}\,\pi^{\frac{i\lambda}{2}}\,\frac{\Gamma(\tau+\frac{1-i\lambda}{2})\,\Gamma(\tau+\frac{1+i\lambda}{2})}{\Gamma(\tau+1)\,\Gamma(\frac{i\lambda}{2})}$$

$$\times (ab)^{-\frac{\tau}{2}}\,|a-b|^{-\frac{1}{2}+\frac{i\lambda}{2}}\,e^{2i\pi(b-a)s}\,\mathfrak{P}^{-\tau}_{-\frac{1}{2}+\frac{i\lambda}{2}}\left(\frac{a+b}{|a-b|}\right).\tag{7.34}$$

Using (7.7), we thus obtain, if $b \neq a$,

$$(\tilde{h}^{\mathrm{iso}})^{\flat}_{\lambda}(s) = 2^{-\frac{3}{2}-\frac{i\lambda}{2}}\,\pi^{-\frac{1}{2}}\,\frac{\Gamma(\tau+\frac{1+i\lambda}{2})}{\Gamma(\frac{i\lambda}{2})}$$

$$\times (ab)^{-\frac{\tau}{2}}\,|a-b|^{-\frac{1}{2}+\frac{i\lambda}{2}}\,e^{2i\pi(b-a)s}\,\mathfrak{P}^{-\tau}_{-\frac{1}{2}+\frac{i\lambda}{2}}\left(\frac{a+b}{|a-b|}\right).\tag{7.35}$$

As a consequence of (7.29), one then has

$$\left(h^{\mathrm{iso}}\right)^{b}_{\lambda}(s) = 2^{-\frac{3}{2}-\frac{i\lambda}{2}} \, \pi^{-\frac{1}{2}} \frac{\Gamma(\tau + \frac{1+i\lambda}{2})}{\Gamma(\frac{i\lambda}{2})} |s|^{-1-i\lambda}$$

$$\times \, (ab)^{-\frac{\tau}{2}} |a-b|^{-\frac{1}{2}+\frac{i\lambda}{2}} \, e^{-\frac{2i\pi(b-a)}{s}} \, \mathfrak{P}^{-\tau}_{-\frac{1}{2}+\frac{i\lambda}{2}} \left(\frac{a+b}{|a-b|}\right). \quad (7.36)$$

We introduce still another species of symbol (the last one except for its dual species, which will not necessitate any new calculation), to wit the soft horocyclic symbol: the isometry property is lost, but the symbol still lives on $\mathbb{R}^2$ and the covariance is preserved. This species of symbol has the following advantage (shared by other types of symbols on $\mathbb{R}^2$, but not by the isometric horocyclic symbol, unless $\tau = \frac{1}{2}$): it satisfies a symmetry property (changing to its negative under $\mathcal{G}$) which is independent of τ, hereafter referred to under the unpleasant vocable of being $\mathcal{G}$-antiinvariant. As an even more decisive bonus, the soft horocyclic symbol of $Q^{\tau}_{a,b}$, not only its homogeneous components, can be fully computed.

Definition 7.3. For $\tau > -1$, the soft horocyclic symbol of an operator with isometric horocyclic symbol h^{iso} is defined as the function h^{soft},

$$h^{\mathrm{soft}} = \frac{2^{-\tau}}{\Gamma(\tau+\frac{3}{2})} \, \Gamma(1 - i\pi\,\mathcal{E}) \, \Gamma\left(\tau + \frac{1}{2} + i\pi\,\mathcal{E}\right) h^{\mathrm{iso}}. \quad (7.37)$$

Observe, as a consequence of Theorem 7.2, that the soft horocyclic symbol is a $\mathcal{G}$-antiinvariant even function (or distribution).

Theorem 7.4. For $\tau > -\frac{1}{2}$, and $a \neq b$, the soft horocyclic symbol of $Q^{\tau}_{a,b}$ is

$$h^{\mathrm{soft}}(x, \xi) = e^{2i\pi(a-b)\frac{\xi}{x}} |x|^{2\tau} \left[x^4 - \frac{(a-b)^2}{4} \right]$$

$$\times \left[\left(x^2 + \frac{1}{2}\left(\sqrt{a}-\sqrt{b}\right)^2\right)\left(x^2 + \frac{1}{2}\left(\sqrt{a}+\sqrt{b}\right)^2\right) \right]^{-\tau-\frac{3}{2}}. \quad (7.38)$$

In the case when $a = b$, one has

$$h^{\mathrm{soft}}(x, \xi) = |x| \, (x^2 + 2a)^{-\tau-\frac{3}{2}} - \delta(x) \int_{-\infty}^{\infty} e^{-4i\pi y\xi} |y| \, (y^2 + 2a)^{-\tau-\frac{3}{2}} \, dy. \quad (7.39)$$

Proof. First consider the case when $a \neq b$. Starting from (7.36) and using (7.11), one obtains

$$h^{\mathrm{soft}}(x, \xi) = \frac{2^{-\tau}}{\Gamma(\tau+\frac{3}{2})} \times 2^{-\frac{3}{2}} \pi^{-\frac{1}{2}} \, (ab)^{-\frac{\tau}{2}} |a-b|^{-\frac{1}{2}} |x|^{-1} \, e^{2i\pi(a-b)\frac{\xi}{x}}$$

$$\int_{-\infty}^{\infty} \frac{i\lambda}{2} \Gamma\left(\tau + \frac{1+i\lambda}{2}\right) \Gamma\left(\tau + \frac{1-i\lambda}{2}\right) \left(\frac{|a-b|}{2\,x^2}\right)^{\frac{i\lambda}{2}} \mathfrak{P}^{-\tau}_{-\frac{1}{2}+\frac{i\lambda}{2}} \left(\frac{a+b}{|a-b|}\right) d\lambda. \quad (7.40)$$

Set

$$e^\beta = \frac{|a-b|}{2\,x^2}\,, \qquad \cosh\alpha = \frac{a+b}{|a-b|}\,, \qquad \sinh\alpha = \frac{2\sqrt{ab}}{|a-b|} : \qquad (7.41)$$

then,

$$\cosh\alpha + \cosh\beta = \frac{1}{|a-b|}\left[a+b+\frac{(a-b)^2}{4\,x^2}+x^2\right]$$

$$= \frac{1}{|a-b|\,x^2}\left(x^2+\frac{1}{2}\left(\sqrt{a}-\sqrt{b}\right)^2\right)\left(x^2+\frac{1}{2}\left(\sqrt{a}+\sqrt{b}\right)^2\right).$$

$$(7.42)$$

Now, one has ([10, p. 205] or [16, p. 409]), if $\tau > -\frac{1}{2}$,

$$\int_{-\infty}^{\infty} e^{\frac{i\beta\lambda}{2}}\,\Gamma\left(\tau+\frac{1+i\lambda}{2}\right)\Gamma\left(\tau+\frac{1-i\lambda}{2}\right)\mathfrak{P}_{-\frac{1}{2}+\frac{i\lambda}{2}}^{-\tau}(\cosh\alpha)\,d\lambda$$

$$= 2^{\frac{3}{2}}\,\pi^{\frac{1}{2}}\,\Gamma\left(\tau+\frac{1}{2}\right)\frac{(\sinh\alpha)^\tau}{(\cosh\alpha+\cosh\beta)^{\tau+\frac{1}{2}}} : \qquad (7.43)$$

note from [16, p. 201] that, as $\lambda \to \infty$, $\mathfrak{P}_{-\frac{1}{2}+\frac{i\lambda}{2}}^{-\tau}(\cosh\alpha) = O(\lambda^{-\tau-1})$, which ensures the convergence of the integral. Then,

$$\left(\Gamma\left(\tau+\frac{3}{2}\right)\right)^{-1}\int_{-\infty}^{\infty}\frac{i\lambda}{2}e^{\frac{i\beta\lambda}{2}}\Gamma\left(\tau+\frac{1+i\lambda}{2}\right)\Gamma\left(\tau+\frac{1-i\lambda}{2}\right)\mathfrak{P}_{-\frac{1}{2}+\frac{i\lambda}{2}}^{-\tau}(\cosh\alpha)\,d\lambda$$

$$= -2^{\frac{3}{2}}\pi^{\frac{1}{2}}\frac{(\sinh\alpha)^\tau\,\sinh\beta}{(\cosh\alpha+\cosh\beta)^{\tau+\frac{3}{2}}}. \qquad (7.44)$$

Theorem 7.4 follows in the case when $a \neq b$. Remaining in this case, we verify that h^{soft} is, indeed, $\mathcal{G}$-antiinvariant. For a general function f on the real line, one has

$$\mathcal{F}\left((x,\xi)\mapsto e^{2i\pi(a-b)\frac{\xi}{x}}\,f(x)\right) = 2\int_{\mathbb{R}^2} e^{2i\pi(a-b)\frac{\eta}{y}}\,f(y)\,e^{4i\pi(x\eta-y\xi)}\,dy$$

$$= 2\int_{-\infty}^{\infty}\delta\left(2x+\frac{a-b}{y}\right)f(y)\,e^{-4i\pi y\xi}\,dy$$

$$= e^{2i\pi(a-b)\frac{\xi}{x}}\,f_1(x) \qquad (7.45)$$

with

$$f_1(x) = \frac{|a-b|}{2x^2}\,f\left(\frac{b-a}{2x}\right). \qquad (7.46)$$

Now, when

$$f(x) = |x|^{2\tau+4}\left[\left(x^2+\frac{1}{2}\left(\sqrt{a}-\sqrt{b}\right)^2\right)\left(x^2+\frac{1}{2}\left(\sqrt{a}+\sqrt{b}\right)^2\right)\right]^{-\tau-\frac{3}{2}}, \qquad (7.47)$$

one has, as easily verified,

$$f_1(x) = \frac{(a-b)^2}{4} |x|^{2\tau} \left[\left(x^2 + \frac{1}{2} \left(\sqrt{a} - \sqrt{b} \right)^2 \right) \left(x^2 + \frac{1}{2} \left(\sqrt{a} + \sqrt{b} \right)^2 \right) \right]^{-\tau - \frac{3}{2}}.$$

$$(7.48)$$

Besides proving that h^{soft} is $\mathcal{G}$-antiinvariant if $a \neq b$, this calculation makes it possible to treat the case when $a = b > 0$ as well. Indeed, the function $(x, \xi) \mapsto e^{2i\pi(a-b)\frac{\xi}{x}} f(x)$ has the obvious limit $x \mapsto |x| (x^2 + 2a)^{-\tau - \frac{3}{2}}$ as $b \to a$. Because of the singularity in the limit at $x = 0$, we could not treat the other term $(x, \xi) \mapsto e^{2i\pi(a-b)\frac{\xi}{x}} f_1(x)$ in the same way. However, there is a limit in the space $\mathcal{S}'(\mathbb{R}^2)$, to wit the transform under $\mathcal{G}$ of the function $(x, \xi) \mapsto |x| (x^2 + 2a)^{-\tau - \frac{3}{2}}$, i.e., the distribution

$$\delta(x) \int_{-\infty}^{\infty} e^{-4i\pi y\xi} |y| (y^2 + 2a)^{-\tau - \frac{3}{2}} dy :$$

$$(7.49)$$

as it turns out, the Fourier integral which is the second factor is a rather complicated special function.   □

As a final effort in this section, we need to define, in connection with the map from an operator to its soft horocyclic symbol, a map in the reverse direction. Given a function K on $\mathbb{R}^2$, set

$$A = \int_0^\infty \int_0^\infty K(b, a) Q_{a,b}^\tau \, da \, db :$$

$$(7.50)$$

then,

$$(Au)(s) = \int_0^\infty \int_0^\infty K(b, a) a^{-\tau} u(a) \delta(s - b) \, da \, db$$

$$= \int_0^\infty \int_0^\infty K(s, a) a^{-\tau} u(a) \, da,$$

$$(7.51)$$

so that K is the integral kernel of A with respect to the measure $dm(s) = s^{-\tau} ds$. From Theorem 7.4, it follows that the soft symbol of A is the function h,

$$h(x, \xi) = \int_0^\infty \int_0^\infty K(b, a) e^{2i\pi(a-b)\frac{\xi}{x}} |x|^{2\tau} \left[x^4 - \frac{(a-b)^2}{4} \right]$$

$$\left[\left(x^2 + \frac{1}{2} \left(\sqrt{a} - \sqrt{b} \right)^2 \right) \left(x^2 + \frac{1}{2} \left(\sqrt{a} + \sqrt{b} \right)^2 \right) \right]^{-\tau - \frac{3}{2}} da \, db. \quad (7.52)$$

Call $\mathcal{B}$ the map: $K \mapsto h$ so defined, and define an adjoint map $\mathcal{B}^*$ by the equation

$$(h \mid \mathcal{B} K)_{L^2(\mathbb{R}^2, \, dx \, d\xi)} = (\mathcal{B}^* h \mid K)_{L^2((0, \infty) \times (0, \infty), \, dm \otimes dm)} .$$

$$(7.53)$$

One finds

$$\overline{(\mathcal{B}^* h)(b, a)} = (ab)^\tau \int_{\mathbb{R}^2} \overline{h}(x, \xi) \, e^{2i\pi(a-b)\frac{\xi}{x}} \, |x|^{2\tau} \left[x^4 - \frac{(a-b)^2}{4} \right.$$

$$\left[\left(x^2 + \frac{1}{2} \left(\sqrt{a} - \sqrt{b} \right)^2 \right) \left(x^2 + \frac{1}{2} \left(\sqrt{a} + \sqrt{b} \right)^2 \right) \right]^{-\tau - \frac{3}{2}} dx \, d\xi, \quad (7.54)$$

hence

$$(\mathcal{B}^* h)(a, b) = (ab)^\tau \int_{\mathbb{R}^2} h(x, \xi) \, e^{2i\pi(a-b)\frac{\xi}{x}} \, |x|^{2\tau} \left[x^4 - \frac{(a-b)^2}{4} \right.$$

$$\left[\left(x^2 + \frac{1}{2} \left(\sqrt{a} - \sqrt{b} \right)^2 \right) \left(x^2 + \frac{1}{2} \left(\sqrt{a} + \sqrt{b} \right)^2 \right) \right]^{-\tau - \frac{3}{2}} dx \, d\xi. \quad (7.55)$$

We now introduce the calculus $\mathrm{Op}_{\mathrm{soft}}^\tau$.

Definition 7.5. Assume $\tau > -\frac{1}{2}$. Given a symbol $h \in \mathcal{S}(\mathbb{R}^2)$, even and $\mathcal{G}$-antiinvariant, the operator $\mathrm{Op}_{\mathrm{soft}}^\tau(h)$ is defined on $C_0^\infty(]0, \infty[)$ by the equation

$$(\mathrm{Op}_{\mathrm{soft}}^\tau(h) \, u)(a) = \int_0^\infty (\mathcal{B}^* h)(a, b) \, u(b) \, b^{-\tau} \, db, \quad (7.56)$$

or

$$(\mathrm{Op}_{\mathrm{soft}}^\tau(h) \, u)(a) = a^\tau \int_0^\infty u(b) \, db \int_{\mathbb{R}^2} h(q, p) \, e^{2i\pi(a-b)\frac{p}{q}} \, |q|^{2\tau} \left[q^4 - \frac{(a-b)^2}{4} \right.$$

$$\left[\left(q^2 + \frac{1}{2} \left(\sqrt{a} - \sqrt{b} \right)^2 \right) \left(q^2 + \frac{1}{2} \left(\sqrt{a} + \sqrt{b} \right)^2 \right) \right]^{-\tau - \frac{3}{2}} dq \, dp. \quad (7.57)$$

Remark 7.1. (i) The operator $\mathrm{Op}_{\mathrm{soft}}^\tau(h)$, as defined by this equation, is still meaningful even if h fails to be $\mathcal{G}$-antiinvariant: however, $\mathcal{G}$-invariant functions in $\mathcal{S}(\mathbb{R}^2)$ will then lie in the nullspace of $\mathrm{Op}_{\mathrm{soft}}^\tau$ if the definition is so extended.

(ii) In the case when h is a radial function, $h(q, p) = H(q^2 + p^2)$, or the transform of a radial function by a linear change of coordinates, the definition is meaningful as soon as $\tau > -1$: under such an extended definition, it will then be possible to assume that $\tau > -1$ rather than $\tau > -\frac{1}{2}$ in Theorems 10.1 and 11.5.

On one hand, the norm in $L^2(\mathbb{R}^2, dm \otimes dm)$ of the integral kernel K above coincides with the Hilbert-Schmidt norm of the operator A as an operator in $L^2(\mathbb{R}, dm)$; on the other hand, this latter norm also coincides with the norm in $L^2(\mathbb{R}^2, dx \, d\xi)$ of the isometric horocyclic symbol of A. It thus follows from (7.53) that the operator giving the isometric horocyclic symbol h^{iso} of $A = \mathrm{Op}_{\mathrm{soft}}^\tau(h)$ in terms of the symbol h is just the adjoint, in $L^2(\mathbb{R}^2, dx \, d\xi)$, of the operator, introduced in Definition 7.3, giving the soft horocyclic symbol in terms of the isometric horocyclic symbol.

One may regard h, as it occurs in Definition 7.5, as still another species of symbol of A, to wit the dual of the soft symbol in the sense given in the introduction of the present chapter: it can be called a *hard* symbol of A, and does not necessitate new calculations. Of course, in view of the size of the Gamma function near infinity on vertical lines, an operator such as $\mathrm{Op}_{\mathrm{soft}}^{\tau}(h)$ with a decent hard symbol h is a *very* nice operator, which is the reason why we have denoted the corresponding map from symbols to operators as $\mathrm{Op}_{\mathrm{soft}}^{\tau}$. We still consider the map from an isometric horocyclic symbol to the corresponding operator as being the main one, to be denoted as Op^{τ}: but using hard symbols will be especially useful in the "arithmetic" Section 10.

Of course, the covariance holds with all species of symbols considered in the present work. The definition of the τ-calculus may look, at first sight, *much* more complicated than the usual Weyl calculus, as (7.57) involves three integrations rather than two: it is admittedly more complicated but, when viewed with the help of the appropriate concept of Wigner function, it demands just the same number of integrations as the usual calculus. The Wigner function $W^{\tau}(v, u)$, u and v lying in $C_0^{\infty}(]0, \infty[)$, is defined so that one should have

$$(v \,|\, \mathrm{Op}_{\mathrm{soft}}^{\tau}(h)\, u)_{\tau+1} = \int_{\mathbb{R}^2} h(x, \xi)\, W^{\tau}(v, u)(x, \xi)\, dx\, d\xi \qquad (7.58)$$

for every appropriate symbol h. It can be computed from the equation (an immediate consequence of (7.56) and (7.55))

$$(v \,|\, \mathrm{Op}_{\mathrm{soft}}^{\tau}(h)\, u)_{\tau+1} = \int_{\mathbb{R}^2} h(x, \xi)\, dx\, d\xi \int_{\mathbb{R}^2} \bar{v}(a)\, u(b)\, e^{2i\pi(a-b)\frac{\xi}{x}}\, |x|^{2\tau}$$
$$\left[x^4 - \frac{(a-b)^2}{4} \right] \left[\left(x^2 + \frac{1}{2}\left(\sqrt{a} - \sqrt{b}\right)^2 \right) \left(x^2 + \frac{1}{2}\left(\sqrt{a} + \sqrt{b}\right)^2 \right) \right]^{-\tau - \frac{3}{2}} da\, db.$$
$$(7.59)$$

Recalling that $P_{v,u} w = (v\,|\,w)\,u$, we may rewrite the left-hand side of (7.58) as $\mathrm{Tr}\,(\mathrm{Op}_{\mathrm{soft}}^{\tau}(h)\, P_{v,u})$. In view of (7.20), this leads to a definition of $\mathrm{Tr}\,(\mathrm{Op}_{\mathrm{soft}}^{\tau}(h)\, Q_{a,b}^{\tau})$, in the case when $h \in \mathcal{S}(\mathbb{R}^2)$, so as to have

$$\mathrm{Tr}\,(\mathrm{Op}_{\mathrm{soft}}^{\tau}(h)\, P_{v,u}) = \int_0^{\infty} \int_0^{\infty} \bar{v}(a)\, u(b)\, \mathrm{Tr}\,(\mathrm{Op}_{\mathrm{soft}}^{\tau}(h)\, Q_{a,b}^{\tau})\, da\, db \qquad (7.60)$$

for every pair of functions $u, v \in H_{\tau+1}$: note, however, that this would not be sufficient to characterize $\mathrm{Tr}\,(\mathrm{Op}_{\mathrm{soft}}^{\tau}(h)\, Q_{a,b}^{\tau})$, since this function of (a, b) would only be defined almost everywhere. To ensure uniqueness, we shall demand, moreover, that this function be continuous. This is important since, in the arithmetic Section 10, we shall indeed have to substitute discretely supported measures for u and v. It leads to the following:

Theorem 7.6. *Assume $\tau > -\frac{1}{2}$. If $a > 0, b > 0$, one can uniquely define $W^\tau(\delta_a, \delta_b)$, as a continuous function of (a, b) with values in $\mathcal{S}'(\mathbb{R}^2)$, so that the following two conditions hold:*

(i) *equation (7.60) is valid for every $\mathcal{G}$-antiinvariant symbol $h \in \mathcal{S}(\mathbb{R}^2)$ and every pair of functions $u, v \in H_{\tau+1}$;*

(ii) *for every pair (a, b) of positive numbers, the distribution $W^\tau(\delta_a, \delta_b)$ is itself $\mathcal{G}$-antiinvariant.*

Explicitly, one has

$$W^\tau(\delta_a, \delta_b)(x, \xi) = e^{2i\pi(a-b)\frac{\xi}{x}} |x|^{2\tau} \left[x^4 - \frac{(a-b)^2}{4} \right]$$

$$\times \left[\left(x^2 + \frac{1}{2} \left(\sqrt{a} - \sqrt{b} \right)^2 \right) \left(x^2 + \frac{1}{2} \left(\sqrt{a} + \sqrt{b} \right)^2 \right) \right]^{-\tau - \frac{3}{2}} \tag{7.61}$$

if $a \neq b$, while

$$W^\tau(\delta_a, \delta_a)(x, \xi) = |x| (x^2 + 2a)^{-\tau - \frac{3}{2}} - \delta(x) \int_{-\infty}^{\infty} e^{-4i\pi y\xi} |y| (y^2 + 2a)^{-\tau - \frac{3}{2}}. \tag{7.62}$$

Hence, the Wigner function of the pair (δ_a, δ_b), is nothing but the soft horocyclic symbol of $Q_{a,b}^\tau$.

The following is a generalization of Theorem 2.1, the even (*resp.* odd) part of which reduces to the case when $\tau = -\frac{1}{2}$ (*resp.* $\tau = \frac{1}{2}$) of the present theorem.

Theorem 7.7. *Assume $\tau > -1$. Let $A = \mathrm{Op}^\tau(h^{\mathrm{iso}})$ be a linear operator on $H_{\tau+1}$ with an isometric horocyclic symbol $h^{\mathrm{iso}} \in \mathcal{S}(\mathbb{R}^2)$, invariant under the symmetry (7.5) (as a horocyclic symbol should be). If one sets $(\mathcal{C} h^{\mathrm{iso}})(z) = (\psi_z^{\tau+1} | A \psi_z^{\tau+1})$, one has*

$$\| \mathcal{C} h^{\mathrm{iso}} \|_{L^2(\Pi)} = \frac{2\pi^{\frac{1}{2}}}{|\Gamma(\tau+1)|} \left\| \Gamma \left(\tau + \frac{1}{2} + i\pi\mathcal{E} \right) h^{\mathrm{iso}} \right\|_{L^2(\mathbb{R}^2)}. \tag{7.63}$$

Proof. From (7.4) and the duplication formula of the Gamma function [16, p. 3], one finds that

$$(2\pi)^{-i\pi\mathcal{E}} \Gamma \left(\tau + \frac{1}{2} + i\pi\mathcal{E} \right) h^{\mathrm{iso}} = 2^\tau \Gamma \left(\frac{\tau}{2} + \frac{1}{4} + \frac{i\pi\mathcal{E}}{2} \right) \Gamma \left(\frac{\tau}{2} + \frac{1}{4} - \frac{i\pi\mathcal{E}}{2} \right) TV f^{\mathrm{act}}. \tag{7.64}$$

Since the operator on the right-hand side, in front of $TV f^{\mathrm{act}}$, is invariant under the change of $i\pi\mathcal{E}$ to its negative, the right-hand side also belongs to the image of TV, a consequence of a property of the Radon transformation: now, on the image of TV, the operator V^*T^* is an isometry [30, p. 27] and a left inverse of TV.

Since the operator $(2\pi)^{-i\pi\mathcal{E}}$ in $L^2(\mathbb{R}^2)$ is unitary, one has

$$
\left\| \Gamma\left(\tau + \frac{1}{2} + i\pi\mathcal{E}\right) h^{\mathrm{iso}} \right\|_{L^2(\mathbb{R}^2)}
$$
$$
= \left\| V^* T^* \left(2^\tau\, \Gamma\left(\frac{\tau}{2} + \frac{1}{4} + \frac{i\pi\mathcal{E}}{2}\right) \Gamma\left(\frac{\tau}{2} + \frac{1}{4} - \frac{i\pi\mathcal{E}}{2}\right) TV\, f^{\mathrm{act}} \right) \right\|_{L^2(\Pi)} ,
$$
$$
\tag{7.65}
$$

which is the same (again, in view of the fact that the conjugate, under the Radon transformation V, of the operator $\Delta - \frac{1}{4}$ is the operator $\pi^2\,\mathcal{E}^2$) as

$$
\left\| 2^\tau\, \Gamma\left(\frac{\tau}{2} + \frac{1}{4} + \frac{i}{2}\sqrt{\Delta - \frac{1}{4}}\right) \Gamma\left(\frac{\tau}{2} + \frac{1}{4} - \frac{i}{2}\sqrt{\Delta - \frac{1}{4}}\right) f^{\mathrm{act}} \right\|_{L^2(\Pi)} . \tag{7.66}
$$

Now, the operator which transforms the active symbol of some operator into its Berezin-covariant symbol is the product of the two operators in (6.31) and in (6.32), in other words

$$
f^{\mathrm{cov}} = \frac{2^{\tau+1}\pi^{\frac{1}{2}}}{\Gamma(\tau+1)}\, \Gamma\left(\frac{\tau}{2} + \frac{1}{4} + \frac{i}{2}\sqrt{\Delta - \frac{1}{4}}\right) \Gamma\left(\frac{\tau}{2} + \frac{1}{4} - \frac{i}{2}\sqrt{\Delta - \frac{1}{4}}\right) f^{\mathrm{act}} .
$$
$$
\tag{7.67}
$$

Theorem 7.7 follows.   □

Chapter 3

Quantization and Modular Forms

We here extend the analysis of operators by means of their diagonal matrix elements, this time against a family of arithmetic coherent states adapted to the representation $\mathcal{D}_{\tau+1}$. The basic distribution $\mathfrak{s}_\tau$ they are built from, substituting for the distributions $\mathfrak{d}_{\text{even}}$ and $\mathfrak{d}_{\text{odd}}$ of Chapter 1, is just another realization of a modular form f of weight $\tau + 1$ of some kind. Section 9 describes some possibilities: one may for instance consider a power of the Dedekind eta-function. The distributions $\mathfrak{s}_\tau^{\tilde{g}}$, where $\tilde{g}$ lies in some homogeneous space of the universal cover of G above $\Gamma\backslash G$, then constitute a family of (arithmetic) coherent states for the representation $\pi_{\tau+1}$ in the way described by a formula of resolution of the identity analogous to (1.2): note that the existence of such a formula depends in a crucial way on the fact that f is a cusp-form.

Let A be an operator with a radial horocyclic symbol h. The family of diagonal matrix elements of A against the transforms of the distribution $\mathfrak{s}^{\tilde{g}}$ under $\mathcal{D}_{\tau+1}$ constitutes, again, an automorphic function $z \mapsto (\mathcal{A}_\tau h)(z)$: the analysis of its Roelcke-Selberg decomposition makes up the (rather technical) Section 10. As a preparation, it has been found necessary, in Section 8, to extend the Rankin-Selberg unfolding method, in a way substituting for the usual construction of Eisenstein's series another one, based on the use of the function $y^{\frac{1}{2}} K_{s-\frac{1}{2}}(2\pi k y)\, e^{2i\pi k x}$ in place of the function y^s: the most natural way to cope with the divergences turns out to be the exact one needed for the application we have in mind. The spectral decomposition of the function $\mathcal{A}_\tau h$ finally involves the convolution L-function $L(\bar{f} \otimes f, s)$ associated with the modular form f. It would be useful to understand more about this function, aside from the special values of τ and special modular forms f for which matters seem to be clear.

It is not necessary to consider only radial symbols: only, if one wants to do away with this assumption, one has to switch from automorphic functions to automorphic distributions, a topic with which the reader is probably not familiar; this is the object of Section 11.

8   An extension of the Rankin-Selberg unfolding method

In view of a crucial application in Section 10, we need to extend here the Rankin-Selberg unfolding method. Let us start by recalling the usual result, in its most standard version, following [4, p. 70] or [37, p. 268].

Proposition 8.1. *Let F be the usual fundamental domain of Γ, to wit $\{z = x + iy\colon |z| > 1, -\frac{1}{2} < x < \frac{1}{2}\}$, and let $\mathcal{D}$ be the strip of Π defined by the sole condition $-\frac{1}{2} < x < \frac{1}{2}$. Let ϕ be a Γ-automorphic C^∞ function, rapidly decreasing at infinity in F. Then, for $\operatorname{Re} s > 1$, one has*

$$\int_{\mathcal{D}} y^s \, \phi(z) \, d\mu(z) = \int_F E(z, s) \, \phi(z) \, d\mu(z) : \tag{8.1}$$

recall that $d\mu(z) = y^{-2} \, dx \, dy$.

Proof. The integral on the left-hand side is convergent because the function ϕ is bounded in F and automorphic, hence bounded in the whole of Π. Let $\Gamma_\infty^o \subset \Gamma$ consist of all matrices $\left(\begin{smallmatrix} 1 & n \\ 0 & 1 \end{smallmatrix}\right)$ with $n \in \mathbb{Z}$, and let Γ_∞ denote the group of all matrices $\pm g$, $g \in \Gamma_\infty^o$. Then $\mathcal{D}$ is exactly covered (up to a one-dimensional subset, a union of transforms of the boundary of F) by the family of domains $g.F$, where g describes an appropriate set of representatives of the quotient set $\Gamma_\infty \backslash \Gamma$. Now, the class of $g = \left(\begin{smallmatrix} a & b \\ c & d \end{smallmatrix}\right) \in \Gamma$ in $\Gamma_\infty^o \backslash \Gamma$ is characterized by the pair c, d, subject to the usual condition $(c, d) = 1$ (as usual, (c, d) denotes the g.c.d. of the pair c, d). On the other hand, if one sets $h_{s,0}(z) = y^s$, the function $h_{s,0}$ is invariant under all fractional-linear transformations associated to matrices in Γ_∞, so that, for $g \in \Gamma$, $h_{s,0} \circ g$ only depends on the class of g in the quotient set $\Gamma_\infty \backslash \Gamma$. This makes it possible to consider, for $\operatorname{Re} s > 1$, the series

$$\sum_{g \in \Gamma_\infty \backslash \Gamma} (h_{s,0} \circ g)(z) = \frac{1}{2} \sum_{\substack{(c,d)=1}} \left(\frac{y}{|cz + d|^2} \right)^s, \tag{8.2}$$

just the definition of $E_s(z, s)$. Finally,

$$\int_{\mathcal{D}} h_{s,0}(z) \, \phi(z) \, d\mu(z) = \sum_{g \in \Gamma_\infty \backslash \Gamma} \int_{g.F} h_{s,0}(z) \, \phi(z) \, d\mu(z)$$

$$= \int_F \sum_{g \in \Gamma_\infty \backslash \Gamma} (h_{s,0} \circ g)(z) \, \phi(z) \, d\mu(z). \tag{8.3}$$

$\square$

Our task in this section is to obtain a comparable result, using the function

$$h_{s,k} : z \mapsto y^{\frac{1}{2}} K_{s-\frac{1}{2}}(2\pi k y) \, e^{2i\pi k x} \,, \qquad k > 0 \,, \tag{8.4}$$

which occurs as the kth term in the Fourier expansion of all periodic generalized eigenfunctions of Δ for the eigenvalue $s(1 - s)$, in place of the function $h_{s,0}$. However, the corresponding integral

$$I(s, \phi) = \int_D y^{\frac{1}{2}} K_{s-\frac{1}{2}}(2\pi k y) \, e^{2i\pi k x} \, \phi(z) \, d\mu(z) \tag{8.5}$$

is divergent. On the other side, the Poincaré series $\sum_{g \in \Gamma_\infty \backslash \Gamma} h_{s,k}(g \cdot z)$ does not converge for any value of s.

We shall have to turn around this difficulty in a very specific way, connected to our needs in Section 10. For $s \neq \frac{1}{2}$, one has [16, p. 67]

$$K_{s-\frac{1}{2}}(2\pi k y) = \frac{2\pi k y}{1 - 2s} \left[K_{s-\frac{3}{2}}(2\pi k y) - K_{s+\frac{1}{2}}(2\pi k y) \right] \,. \tag{8.6}$$

Now, as will be seen, the integral (note the exponent of y)

$$A(s, \phi) = \int_D y^{\frac{3}{2}} K_{s-\frac{1}{2}}(2\pi k y) \, e^{2i\pi k x} \, \phi(z) \, d\mu(z) \tag{8.7}$$

is meaningful for s in some non-void domain of the complex plane: to give (8.5) a meaning, it then looks as if one could define it as

$$I(s, \phi) = \frac{2\pi k}{1 - 2s} \left[A(s - 1, \phi) - A(s + 1, \phi) \right] \,. \tag{8.8}$$

Of course, the integrals defining the two terms of this decomposition cannot be simultaneously convergent (since the one defining $I(s, \phi)$ is not): but, as will be seen, each of them has an *analytic extension* to some larger domain, which will finally give the right-hand side of (8.8) a meaning.

Towards constructing the other side of the proposed formula, set

$$c_{s,k}(z) = y^{\frac{3}{2}} K_{s-\frac{1}{2}}(2\pi k y) \, e^{2i\pi k x} \,, \tag{8.9}$$

so that, from (8.6) again,

$$h_{s,k}(z) = \frac{2\pi k}{1 - 2s} \left[c_{s-1,k}(z) - c_{s+1,k}(z) \right] \,. \tag{8.10}$$

These functions will occur later in some integrals: since Gamma factors will consistently show up, up and down, in these calculations, it will help to have some estimate of such functions as $|\text{Im } s| \to \infty$. For $a > 0$, it follows from the expansion [16, p. 66]

$$I_{\sigma-\frac{1}{2}+it}(2a) = \frac{1}{\Gamma(\sigma + \frac{1}{2} + it)}$$

$$\times \sum_{m \geq 0} \frac{a^{\sigma-\frac{1}{2}+it+2m}}{m!\,(\sigma + \frac{1}{2} + it)(\sigma + \frac{1}{2} + it + 1)\dots(\sigma + \frac{1}{2} + it + m - 1)} \quad (8.11)$$

that, provided that $\sigma > -1$,

$$|I_{\sigma-\frac{1}{2}+it}(2a)| \leq \frac{a^{\sigma-\frac{1}{2}}}{|\Gamma(\sigma + \frac{1}{2} + it)|} \exp \frac{a^2}{|\sigma + \frac{1}{2}|} : \quad (8.12)$$

using the equation

$$K_{s-\frac{1}{2}}(2\pi k y) = -\frac{1}{2}\Gamma\left(\frac{1}{2}+s\right)\Gamma\left(\frac{1}{2}-s\right)\left[I_{\frac{1}{2}-s}(2\pi k y) - I_{s-\frac{1}{2}}(2\pi k y)\right], \quad (8.13)$$

one obtains

$$|K_{\sigma-\frac{1}{2}+it}(2\pi k y)| \leq \frac{1}{2}\left|\Gamma\left(\sigma + \frac{1}{2}\cdot 1 + it\right)\Gamma\left(\frac{1}{2} - \sigma - it\right)\right|$$

$$\times \left[\frac{(\pi k y)^{-\sigma}}{|\Gamma(\frac{3}{2} - \sigma - it)|}\exp\frac{\pi^2 k^2 y^2}{|\frac{3}{2} - \sigma|} + \frac{(\pi k y)^{\sigma-1}}{|\Gamma(\sigma + \frac{1}{2} + it)|}\exp\frac{\pi^2 k^2 y^2}{|\sigma + \frac{1}{2}|}\right]. \quad (8.14)$$

Together with the equivalent, as $|t| \to \infty$ [16, p. 13]

$$|\Gamma(\sigma + it)| \sim (2\pi)^{\frac{1}{2}}|t|^{\sigma-\frac{1}{2}}e^{-\frac{\pi|t|}{2}}, \quad (8.15)$$

it shows that some factor, roughly (up to powers of $|t|$) of the size of $e^{-\frac{\pi|t|}{2}}$, is saved in the function $|K_{\sigma-\frac{1}{2}+it}(2\pi k y)|$.

Theorem 8.2. *With $k = 1, 2, \dots$ and $c_{s,k}$ as defined in (8.9), set*

$$\mathfrak{f}_s(z) = \sum_{g \in \Gamma_\infty \backslash \Gamma} c_{s,k}(g \cdot z) \quad (8.16)$$

(again, the function $c_{s,k}$ is invariant under translations $z \mapsto z + 1$), in other words

$$\mathfrak{f}_s(z) = \frac{1}{2}\sum_{(m,n)=1}\left(\frac{y}{|mz+n|^2}\right)^{\frac{3}{2}}K_{s-\frac{1}{2}}\left(\frac{2\pi k y}{|mz+n|^2}\right)$$

$$\times \exp\left(2i\pi k \,\text{Re}\,\frac{az+b}{mz+n}\right), \quad \left(\begin{smallmatrix} a & b \\ m & n \end{smallmatrix}\right) \in \Gamma. \quad (8.17)$$

The series converges when $\frac{1}{2} < \operatorname{Re} s < 1$, and the function $s \mapsto \mathfrak{f}_s(z)$ so defined is holomorphic. It extends as a meromorphic function in the whole complex plane, with two families of poles, all simple: the ones from the first family are located at points $s = n + \frac{3}{2} \pm \frac{i\lambda_j}{2}$ or $s = -n - \frac{1}{2} \pm \frac{i\lambda_j}{2}$ where $n = 0, 1, \dots$ and $\left(\frac{1+\lambda_j^2}{4}\right)$ is the sequence of eigenvalues of the hyperbolic Laplacian Δ in $L^2(\Gamma\backslash\Pi)$; the ones from the second family are to be found within the sequence $\{-\frac{1}{2} - n, \; n = 0, 1, \dots\}$ or $\{\frac{3}{2} + n, \; n = 0, 1, \dots\}$. The function

$$E_k(z, s) = \frac{2\pi k}{1 - 2s} \left[\mathfrak{f}_{s-1}(z) - \mathfrak{f}_{s+1}(z) \right] \tag{8.18}$$

coincides with $\alpha_k(s)\, E(z, s)$ for some function α_k to be determined in Section 10. As a function of s, $E_k(z, s)$ has no singularity on the line $\frac{1}{2} + i\mathbb{R}$.

Proof. It is convenient to use the following set of inequalities, in which d denotes the hyperbolic distance in Π:

$$y\, e^{-d(i,z)} \, (m^2 + n^2) \leq |mz + n|^2 \leq y\, e^{d(i,z)} \, (m^2 + n^2), \tag{8.19}$$

proved in the following way: one has

$$|mz + n|^2 = |z|^2 m^2 + 2x\, mn + n^2 = \left(\begin{smallmatrix} m \\ n \end{smallmatrix}\right)^\top Q \left(\begin{smallmatrix} m \\ n \end{smallmatrix}\right), \tag{8.20}$$

with $Q = \left(\begin{smallmatrix} |z|^2 & x \\ x & 1 \end{smallmatrix}\right)$, a matrix with eigenvalues $\frac{1}{2}\left[1 + |z|^2 \pm \sqrt{(1 + |z|^2)^2 - 4y^2}\,\right]$, i.e., $y \exp(\pm d(i, z))$, if one remembers that $\cosh d(i, z) = \frac{1+|z|^2}{2y}$.

Hence, $\frac{y}{|mz+n|^2}$ is of the order of $\frac{1}{m^2+n^2}$ for z in any compact subset of Π: assuming that $\operatorname{Re} s > \frac{1}{2}$ and $s \notin \frac{1}{2} + \mathbb{Z}$, one has [16, p. 66] $K_{s-\frac{1}{2}}(\varepsilon) \sim C_s \varepsilon^{\frac{1}{2}-s}$ as $\varepsilon \to 0$ for some constant C_s. It follows from (8.17) that, for z in any compact subset of Π, the general term of the series for $\mathfrak{f}_s(z)$ is majorized by $C\, (m^2 + n^2)^{-2+\operatorname{Re} s}$ for some $C > 0$. For $\frac{1}{2} < \operatorname{Re} s < 1$, the function $s \mapsto \mathfrak{f}_s(z)$ is thus well defined and analytic: we need to extend its domain.

To do this, we shall appeal to the non-holomorphic Poincaré-Selberg series introduced in [21], also used in [9, 8] in investigations relative to the Kloosterman sums. For any integer $k > 0$ and $s \in \mathbb{C}$ with $\operatorname{Re} s > 1$, set

$$U_k(z, s) = \frac{1}{2} \sum_{(m,n)=1} \left(\frac{y}{|mz + n|^2}\right)^s \exp\left(2i\pi k\, \frac{az + b}{mz + n}\right), \tag{8.21}$$

where it is still assumed that $\left(\begin{smallmatrix} a & b \\ m & n \end{smallmatrix}\right) \in \Gamma$. The function $U_0(z, s)$, though meaningful, reduces to $E(z, s)$: we discard this case in order to take advantage of the fact that the function $z \mapsto U_k(z, s)$ lies in $L^2(\Gamma\backslash\Pi)$. As shown by Selberg [21], the functions under consideration satisfy the differential equation

$$[\Delta - s(1 - s)]\, U_k(z, s) = 4\pi s\, U_k(z, s + 1), \tag{8.22}$$

and the function $s \mapsto U_k(z, s)$ extends as a meromorphic function in the entire plane. The coefficients of the Roelcke-Selberg decomposition of the function $U_k(\cdot, s)$ are fully explicit, as follows. On one hand [9, p. 247] or [8, p. 246] or [14, p. 406], one has

$$\int_{\Gamma\backslash\Pi} U_k(z, s) \, E\left(z, \frac{1+i\lambda}{2}\right) d\mu(z)$$

$$= \frac{2^{2-2s} \pi^{-s+\frac{3+i\lambda}{2}}}{\zeta(1+i\lambda)} k^{-s+\frac{1-i\lambda}{2}} \sigma_{i\lambda}(k) \frac{\Gamma(s-\frac{1}{2}-\frac{i\lambda}{2})\Gamma(s-\frac{1}{2}+\frac{i\lambda}{2})}{\Gamma(s)\Gamma(\frac{1+i\lambda}{2})}, \quad (8.23)$$

where

$$\sigma_{i\lambda}(k) = \sum_{d\geq 1, \, d|k} d^{i\lambda}. \quad (8.24)$$

On the other hand (*loc. cit.*), if $\mathcal{M}$ is a cusp-form corresponding to some eigenvalue $\frac{1+\lambda_j^2}{4}$ of Δ, with the Fourier expansion

$$\mathcal{M}(z) = y^{\frac{1}{2}} \sum_{k\neq 0} b_k \, K_{\frac{i\lambda_j}{2}}(2\pi|k|y) \, e^{2i\pi kx}, \quad (8.25)$$

one has

$$\int_{\Gamma\backslash\Pi} U_k(z, s) \overline{\mathcal{M}(z)} \, d\mu(z) = \overline{b_k} \times \pi^{\frac{1}{2}} (4\pi k)^{\frac{1}{2}-s} \frac{\Gamma(s-\frac{1}{2}-\frac{i\lambda_j}{2})\Gamma(s-\frac{1}{2}+\frac{i\lambda_j}{2})}{\Gamma(s)}. \quad (8.26)$$

Let $(\mathcal{M}_j)_{j\geq 1}$ be an orthonormal basis of the space of cusp-forms in $L^2(\Gamma\backslash\Pi)$, the corresponding sequence of eigenvalues of Δ being given as $(\frac{1+\lambda_j^2}{4})_{j\geq 1}$ (so that repetitions are possible in the – possibly unlikely – case when the eigenspace corresponding to some eigenvalue had dimension > 1). For every $j \geq 1$, let $(b_k^j)_{k\in\mathbb{Z}^\times}$ be the sequence of Fourier coefficients, as they appear in (8.25), of $\mathcal{M}_j$. From (8.23) and [14, p. 391], the spectral decomposition of $U_k(z, s)$ is given as the sum

$$\frac{(4\pi k)^{-s}}{\Gamma(s)} \int_{-\infty}^{\infty} \frac{(\pi k)^{\frac{1+i\lambda}{2}}}{\zeta(1+i\lambda)} \sigma_{-i\lambda}(k) \frac{\Gamma(s-\frac{1}{2}-\frac{i\lambda}{2})\Gamma(s-\frac{1}{2}+\frac{i\lambda}{2})}{\Gamma(\frac{1+i\lambda}{2})} E\left(z, \frac{1-i\lambda}{2}\right) d\lambda$$

$$+ \frac{\pi^{\frac{1}{2}} (4\pi k)^{\frac{1}{2}-s}}{\Gamma(s)} \sum_{j\geq 1} \Gamma\left(s-\frac{1}{2}-\frac{i\lambda_j}{2}\right) \Gamma\left(s-\frac{1}{2}+\frac{i\lambda_j}{2}\right) \overline{b_k^j} \, \mathcal{M}_j(z). \quad (8.27)$$

As observed by Selberg [21], this spectral decomposition gives the analytic continuation of the function $s \mapsto U_k(z, s)$ as a meromorphic function in the entire complex plane: the poles are simple and located at the points $s = \frac{1}{2} \pm \frac{i\lambda_j}{2} - n$ with $n = 0, 1, \ldots$.

We now express $\mathfrak{f}_s(z)$ (cf. (8.17)), for $\frac{1}{2} < \mathrm{Re}\, s < 1$, as a series of Poincaré-Selberg functions. Using (8.13),

$$
K_{s-\frac{1}{2}}\left(\frac{2\pi k y}{|mz+n|^2}\right) = -\frac{1}{2}\,\Gamma\left(\frac{1}{2}+s\right)\Gamma\left(\frac{1}{2}-s\right) \tag{8.28}
$$

$$
\times\left[\sum_{\alpha\geq 0}\frac{\left(\frac{\pi k y}{|mz+n|^2}\right)^{-s+\frac{1}{2}+2\alpha}}{\alpha!\,\Gamma(-s+\frac{3}{2}+\alpha)} - \sum_{\alpha\geq 0}\frac{\left(\frac{\pi k y}{|mz+n|^2}\right)^{s-\frac{1}{2}+2\alpha}}{\alpha!\,\Gamma(s+\frac{1}{2}+\alpha)}\right].
$$

On the other hand,

$$
\mathrm{Re}\,\frac{az+b}{mz+n} = \frac{az+b}{mz+n} - \frac{i\,y}{|mz+n|^2}, \tag{8.29}
$$

so that

$$
\exp\left(2i\pi k\,\mathrm{Re}\,\frac{az+b}{mz+n}\right) = \exp\left(2i\pi k\,\frac{az+b}{mz+n}\right)\sum_{\beta\geq 0}\frac{(2\pi k)^\beta}{\beta!}\left(\frac{y}{|mz+n|^2}\right)^\beta. \tag{8.30}
$$

Taking advantage of (8.28) and (8.30), we obtain from (8.17) the equation

$$
\mathfrak{f}_s(z) = -\frac{1}{2}\,\Gamma\left(\frac{1}{2}+s\right)\Gamma\left(\frac{1}{2}-s\right)\sum_{\alpha,\beta\geq 0}\frac{(2\pi k)^\beta}{\alpha!\,\beta!}\,[g_{\alpha,\beta;\,s}(z) - g_{\alpha,\beta;\,1-s}(z)] \tag{8.31}
$$

with

$$
g_{\alpha,\beta;\,s}(z) = \frac{(\pi k)^{-s+\frac{1}{2}+2\alpha}}{\Gamma(-s+\frac{3}{2}+\alpha)}\times\frac{1}{2}\sum_{(m,n)=1}\left(\frac{y}{|mz+n|^2}\right)^{-s+2+2\alpha+\beta}\exp\left(2i\pi k\frac{az+b}{mz+n}\right)
$$

$$
= \frac{(\pi k)^{-s+\frac{1}{2}+2\alpha}}{\Gamma(-s+\frac{3}{2}+\alpha)}\,U_k(z,\,-s+2+2\alpha+\beta). \tag{8.32}
$$

Finally, for $\frac{1}{2} < \mathrm{Re}\, s < 1$, one has

$$
\mathfrak{f}_s(z) = -\frac{1}{2}\,\Gamma\left(\frac{1}{2}+s\right)\Gamma\left(\frac{1}{2}-s\right)\sum_{\alpha,\beta\geq 0}\frac{(2\pi k)^\beta}{\alpha!\,\beta!} \tag{8.33}
$$

$$
\times\left[\frac{(\pi k)^{-s+\frac{1}{2}+2\alpha}}{\Gamma(-s+\frac{3}{2}+\alpha)}\,U_k(z,-s+2+2\alpha+\beta) - \frac{(\pi k)^{s-\frac{1}{2}+2\alpha}}{\Gamma(s+\frac{1}{2}+\alpha)}\,U_k(z,s+1+2\alpha+\beta)\right].
$$

From (8.21), the convergence of the series defining $U_k(z, s)$ when $\mathrm{Re}\, s > 1$ only improves when s is replaced by $s+1$, which ensures the convergence of the series in (8.33). This equation thus provides the analytic continuation of the function

$f_s(z)$ as a meromorphic function of s in the whole complex plane: the first family of poles, as listed in the statement of Theorem 8.2, is obtained from Selberg's analysis (as above recalled) of the functions $U_k(z, s)$. There is a second family of (simple) poles, which originates from the two Gamma factors in front of the right-hand side of (8.33). At first sight, it would look as if all elements of the sequence $\frac{1}{2} + \mathbb{Z}$ might do, but it is important for the application in Section 10 to note that $s = \frac{1}{2}$ is to be excluded from the list: indeed, for this value of s, the two terms within the bracket on the right-hand side cancel off.

For every $s \in \mathbb{C}$ not a pole of the function $s \mapsto E_k(z, s)$, the function $E_k(\bullet, s)$ is automorphic: we next show that it satisfies the eigenvalue equation

$$\Delta E_k(z, s) = s(1 - s) E_k(z, s). \tag{8.34}$$

Starting from (8.9), one has

$$[\Delta - s(1 - s)] c_{s,k}(z) = -y^{\frac{3}{2}} e^{2i\pi kx} \left[K_{s-\frac{1}{2}}(2\pi ky) + 4\pi ky K'_{s-\frac{1}{2}}(2\pi ky) \right] \tag{8.35}$$

or, using [16, p. 66]

$$K_\nu(z) + 2z K'_\nu(z) = z \left[\left(\frac{1}{2\nu} - 1 \right) K_{\nu+1}(z) - \left(\frac{1}{2\nu} + 1 \right) K_{1-\nu}(z) \right], \tag{8.36}$$

$$[\Delta - s(1 - s)] c_{s,k}(z)$$
$$= 2\pi k \, y^{\frac{5}{2}} e^{2i\pi kx} \left[\frac{2s - 2}{2s - 1} K_{s+\frac{1}{2}}(2\pi ky) + \frac{2s}{2s - 1} K_{\frac{3}{2}-s}(2\pi ky) \right]. \tag{8.37}$$

Besides the function $c_{s,k}$, let us introduce the function

$$c^1_{s,k}(z) = 2\pi k \, y^{\frac{5}{2}} K_{s-\frac{1}{2}}(2\pi ky) \, e^{2i\pi kx} \tag{8.38}$$

and define accordingly

$$f^1_s(z) = \sum_{g \in \Gamma_\infty \backslash \Gamma} c^1_{s,k}(g \cdot z). \tag{8.39}$$

The series for $f^1_s(z)$ is convergent for $\frac{1}{2} < \mathrm{Re}\, s < 2$, and one has if $\frac{1}{2} < \mathrm{Re}\, s < 1$, in which case $s + 1$ and $2 - s$ each have a real part between $\frac{1}{2}$ and 2,

$$[\Delta - s(1 - s)] f_s = \frac{2s - 2}{2s - 1} f^1_{s+1} + \frac{2s}{2s - 1} f^1_{2-s}. \tag{8.40}$$

Still under the same assumptions, one has

$$f^1_{2-s} - f^1_{s+1} = (1 - 2s) f_s, \tag{8.41}$$

as it follows simply from [16, p. 67]

$$z \left[K_{\frac{3}{2}-s}(z) - K_{s+\frac{1}{2}}(z) \right] = (1 - 2s) K_{s-\frac{1}{2}}(z) \tag{8.42}$$

together with the expansions (8.17) and (8.39), valid for s in this range of values. Also, the series (8.17) for $\mathfrak{f}_s(z)$ becomes that for $\mathfrak{f}_s^1(z)$ provided one just inserts an extra factor $\frac{2\pi ky}{|mz+n|^2}$: as a consequence, (8.33) transforms to

$$\mathfrak{f}_s^1(z) = -\frac{1}{2}\Gamma\left(\frac{1}{2}+s\right)\Gamma\left(\frac{1}{2}-s\right)\sum_{\alpha,\beta\geq 0}\frac{(2\pi k)^{\beta+1}}{\alpha!\beta!} \tag{8.43}$$

$$\times\left[\frac{(\pi k)^{-s+\frac{1}{2}+2\alpha}}{\Gamma(-s+\frac{3}{2}+\alpha)}U_k(z,-s+3+2\alpha+\beta) - \frac{(\pi k)^{s-\frac{1}{2}+2\alpha}}{\Gamma(s+\frac{1}{2}+\alpha)}U_k(z,s+2+2\alpha+\beta)\right].$$

Again, the series (8.43) provides the analytic continuation of the function $s\mapsto \mathfrak{f}_s^1(z)$ as a meromorphic function in the entire complex plane.

From (8.33) and (8.43), one has

$$\mathfrak{f}_s = \mathfrak{f}_{1-s}, \qquad \mathfrak{f}_s^1 = \mathfrak{f}_{1-s}^1 \tag{8.44}$$

(not a surprising fact from the equation $K_{s-\frac{1}{2}} = K_{\frac{1}{2}-s}$: however, it could not be proved in this way). Rewriting (8.41) as

$$\mathfrak{f}_{s-1}^1 - \mathfrak{f}_{s+1}^1 = (1-2s)\,\mathfrak{f}_s \tag{8.45}$$

and applying it twice (once with $s-1$, once with $s+1$ in place of s), one can verify that

$$\left(\frac{2s-4}{2s-3} - \frac{2s+2}{2s+1}\right)\mathfrak{f}_s^1 + \frac{2s-2}{2s-3}\,\mathfrak{f}_{s-2}^1 - \frac{2s}{2s+1}\,\mathfrak{f}_{s+2}^1$$

$$= \frac{2s}{2s+1}\,(\mathfrak{f}_s^1 - \mathfrak{f}_{s+2}^1) + \frac{2s-2}{2s-3}\,(\mathfrak{f}_{s-2}^1 - \mathfrak{f}_s^1)$$

$$= (2-2s)\,\mathfrak{f}_{s-1} - 2s\,\mathfrak{f}_{s+1}. \tag{8.46}$$

Rewriting now (8.40) as

$$[\Delta - s(1-s)]\,\mathfrak{f}_s = \frac{2s-2}{2s-1}\,\mathfrak{f}_{s+1}^1 + \frac{2s}{2s-1}\,\mathfrak{f}_{s-1}^1, \tag{8.47}$$

so that

$$[\Delta - s(1-s)]\,\mathfrak{f}_{s-1} = \frac{2s-4}{2s-3}\,\mathfrak{f}_s^1 + \frac{2s-2}{2s-3}\,\mathfrak{f}_{s-2}^1 + (2s-2)\,\mathfrak{f}_{s-1},$$

$$[\Delta - s(1-s)]\,\mathfrak{f}_{s+1} = \frac{2s}{2s+1}\,\mathfrak{f}_{s+2}^1 + \frac{2s+2}{2s+1}\,\mathfrak{f}_s^1 - 2s\,\mathfrak{f}_{s+1}, \tag{8.48}$$

and applying (8.18) and (8.46), we obtain

$$\Delta\,E_k(z,s) = s(1-s)\,E_k(z,s). \tag{8.49}$$

This proves (8.18) since the function $E_k(\cdot,s)$ is also automorphic.

It remains to be seen that $E_k(z, s)$ has no singularity on the line $\mathrm{Re} = \frac{1}{2}$, showing to begin with that a point such as $\frac{1}{2} + \frac{i\mu}{2}$, with $\frac{1+\mu^2}{4}$ in the discrete spectrum of Δ, is a regular value of $E_k(z, s)$. We start from the expansion (8.27) of $U_k(z, s)$: first, recalling that $(b_k^j)_{k \in \mathbb{Z}^\times}$ is the sequence of Fourier coefficients of $\mathcal{M}_j$, define

$$\mathcal{N}_k^\mu(z) = \sum \overline{b_k^j}\, \mathcal{M}_j(z)\,, \tag{8.50}$$

where the sum is extended to all j's such that $\lambda_j^2 = \mu^2$. From the expansion under review, the residue of $U_k(z, s)$ at $s = \frac{1}{2} + \frac{i\mu}{2} - n$ (with $n = 0, 1, \dots$) is

$$\frac{\pi^{\frac{1}{2}}\, (4\pi k)^{n - \frac{i\mu}{2}}}{\Gamma(\frac{1}{2} + \frac{i\mu}{2} - n)}\, \frac{(-1)^n}{n!}\, \Gamma(-n + i\mu)\, \mathcal{N}_k^\mu(z)\,. \tag{8.51}$$

In particular, using the duplication formula of the Gamma function, one can also write

$$\mathrm{Res}\left[U_k(z, s);\, \frac{1}{2} + \frac{i\mu}{2}\right] = \frac{1}{2}\, (\pi k)^{-\frac{i\mu}{2}}\, \Gamma\left(\frac{i\mu}{2}\right) \mathcal{N}_k^\mu(z)\,. \tag{8.52}$$

We now use (8.33) and interest ourselves only in the poles of $\mathfrak{f}_s(z)$ with a real part closest from $\frac{1}{2}$ i.e., $-\frac{1}{2}$ or $\frac{3}{2}$. From (8.51), we obtain

$$\mathrm{Res}\left[\mathfrak{f}_s(z);\, \frac{3}{2} - \frac{i\mu}{2}\right] = \frac{1}{4\pi k}\, \Gamma\left(2 - \frac{i\mu}{2}\right) \Gamma\left(-1 + \frac{i\mu}{2}\right) \mathcal{N}_k^\mu(z)\,,$$

$$\mathrm{Res}\left[\mathfrak{f}_s(z);\, -\frac{1}{2} + \frac{i\mu}{2}\right] = \frac{1}{4\pi k}\, \Gamma\left(\frac{i\mu}{2}\right) \Gamma\left(1 - \frac{i\mu}{2}\right) \mathcal{N}_k^\mu(z) \tag{8.53}$$

(as a safeguard, one residue is the negative of the other, as necessary from the equation $\mathfrak{f}_s = \mathfrak{f}_{1-s}$). Then,

$$\mathrm{Res}\left[\mathfrak{f}_{s-1}(z) - \mathfrak{f}_{s+1}(z);\, \frac{1}{2} + \frac{i\mu}{2}\right] = \mathrm{Res}\left[\mathfrak{f}_s(z);\, -\frac{1}{2} + \frac{i\mu}{2}\right] - \mathrm{Res}\left[\mathfrak{f}_s(z);\, \frac{3}{2} + \frac{i\mu}{2}\right] \tag{8.54}$$

is zero, which almost concludes the proof of Theorem 8.2. We have not yet examined, however, the poles $-\frac{1}{2}$ or $\frac{3}{2}$ of $\mathfrak{f}_s(z)$. There is actually no need to make the residues there explicit since, in view of (8.44), one has $f_{-\frac{1}{2}} = f_{\frac{3}{2}}$, so that the point $s = \frac{1}{2}$ must be a regular value of $E_k(z, s)$ anyway. $\square$

Remark 8.1. It will be found in Section 10 that

$$\alpha_k(s) = \frac{1}{2}\, \frac{\Gamma(s - \frac{1}{2})\Gamma(\frac{1}{2} - s)}{\zeta^*(2 - 2s)}\, k^{s - \frac{1}{2}}\, \sigma_{1-2s}(k)\,, \tag{8.55}$$

with $\sigma_{1-2s}(k) = \sum_{1 \le d | k} d^{1-2s}$. Trying to prove this now would lead to an unnecessary task, since this will be obtained without any extra work from the application, there, of the following generalization of the Rankin-Selberg method.

Corollary 8.3. *Let* $k = 1, 2, \ldots$ *and let* ϕ *be a* Γ-*automorphic* C^∞ *function, rapidly decreasing at infinity in the fundamental domain* F. *Recall the definition* (8.9) *of the function* $c_{s,k}$, *and set*

$$A(s, \phi) = \int_{\mathcal{D}} c_{s,k}(z)\, \phi(z)\, d\mu(z), \qquad (8.56)$$

a convergent integral if $\frac{1}{2} < \operatorname{Re} s < 1$ *(recall that* $\mathcal{D}$ *is the strip defined by* $-\frac{1}{2} < \operatorname{Re} z < \frac{1}{2}$*). The function* $s \mapsto A(s, \phi)$ *is holomorphic, and extends as a meromorphic function in the entire complex plane, still denoted in the same way: the poles are simple and contained in those of the function* $s \mapsto \mathfrak{f}_s(z)$ *as made explicit in Theorem 8.2. If neither* $s - 1$ *nor* $s + 1$ *coincides with any of these poles, and* $s \neq \frac{1}{2}$, *one has*

$$\frac{2\pi k}{1 - 2s}\, [A(s - 1, \phi) - A(s + 1, \phi)] = \int_{\Gamma \backslash \Pi} E_k(z, s)\, \phi(z)\, d\mu(z). \qquad (8.57)$$

Proof. Use the usual Rankin-Selberg method to obtain, for s in some appropriate non-void domain,

$$A(s, \phi) = \int_F \mathfrak{f}_s(z)\, \phi(z)\, d\mu(z), \qquad (8.58)$$

and use analytic continuation. □

9 Discrete series and arithmetic coherent states

In the present section, more questions are raised than answered. It seems clear that some understanding of modular forms of general real weight going much beyond the present author's knowledge would help here.

We start from a certain class of modular forms modelled after Dedekind's η-function.

Definition 9.1. Let $\tau > -1$ be given. A τ-adapted distribution $\mathfrak{s}_\tau$ on the line is a measure of the kind

$$\mathfrak{s}_\tau(t) = \sum_{m \geq 0} a_m\, \delta(t - m - \kappa) \qquad (9.1)$$

for some $\kappa > 0$, where the sequence (a_m) is assumed to be controlled by some power of $m + 1$, satisfying the following property. Set

$$f(z) = q^\kappa \sum_{m \geq 0} a_m\, q^m \qquad (9.2)$$

with $q = e^{2i\pi z}$ and $q^\kappa = e^{2i\pi\kappa z}$: it is assumed that the function f is invariant, up to the multiplication by some complex number of absolute value 1, by the transformation $\pi_{\tau+1}\left(\left(\begin{smallmatrix} 0 & 1 \\ -1 & 0 \end{smallmatrix}\right)\right)$ as defined in (6.5). Note that κ becomes unique when subject to the condition $\kappa \leq 1$.

As an example, starting from the η-function, one may take

$$f(z) = (\eta(z))^{2\tau+2} = q^{\frac{\tau+1}{12}} \prod_{n \geq 1} (1 - q^n)^{2\tau+2} \tag{9.3}$$

and, setting

$$f(z) = q^{\frac{\tau+1}{12}} \sum_{m \geq 0} a_m q^m , \tag{9.4}$$

define $\mathfrak{s}_\tau$ as

$$\mathfrak{s}_\tau(t) = \sum_{m \geq 0} a_m \, \delta\left(t - m - \frac{\tau+1}{12}\right) . \tag{9.5}$$

Coming back to the general case, and writing

$$f(z) = \int_0^\infty \mathfrak{s}_\tau(t) \, e^{2i\pi tz} \, dt , \tag{9.6}$$

one observes that, if one forgets about constant factors, and with $v = \mathfrak{s}_\tau$, the map $v \mapsto f$ is the composition of the map $v \mapsto f$ in (6.4) (note that, anyway, the coefficient in front of the integral there would not be meaningful for $\tau \leq 0$) with the map $\pi_{\tau+1}\left(\left(\begin{smallmatrix} 0 & 1 \\ -1 & 0 \end{smallmatrix}\right)\right)$ as made explicit in (6.5) so that, as a consequence of Theorem 6.2, one has

$$\mathcal{D}_{\tau+1}(g) \, \mathfrak{s}_\tau = \omega(g) \, \mathfrak{s}_\tau \tag{9.7}$$

for every $g \in \Gamma$ and some $\omega(g)$ with $|\omega(g)| = 1$. Of course, we have implicitly extended the representation $\pi_{\tau+1}$ to a space of holomorphic functions in Π with a less restricted global behaviour than the ones in $\widetilde{H}_{\tau+1}$, and have used the extension of the representation $\mathcal{D}_{\tau+1}$ to some distribution space, as defined in (7.13).

The function f is rapidly decreasing at infinity in the fundamental domain F of the full modular group. Then,

$$|a_m| \leq C \, m^{\frac{\tau+1}{2}} , \qquad m \geq 1 \tag{9.8}$$

for some constant $C > 0$. This is proved by the usual Hecke argument [22, p. 152] starting from the fact that the function $z \mapsto y^{\frac{\tau+1}{2}} |f(z)|$ is Γ-automorphic, hence bounded not only in the fundamental domain, but in Π. In the example case (9.3), this is not optimal either when $\tau = -\frac{1}{2}$ (3.42) or when $\tau = 11$ (Deligne's theorem, formerly Ramanujan's conjecture).

Other examples of τ-adapted distribution are obtained, in the case when τ is an integer or a half-integer, from the decomposition of a function (or distribution) in $\mathbb{R}^n$, $n = 1, 2, \ldots$, under the group of rotations.

Proposition 9.2. *Let* $n = 1, 2, \ldots,$ $k = 0, 1, \ldots$ *and* $\tau = \frac{n-2}{2} + k$. *For* $\mu = (\mu_1, \ldots, \mu_n) \in \mathbb{Z}^n$, *set*

$$\chi(\mu) = \prod_{j=1}^n \chi^{(12)}(\mu_j) . \tag{9.9}$$

Given any harmonic polynomial P^k, homogeneous of degree k, the distribution

$$\mathfrak{s}_P^k(t) = \sum_{m \geq 0} a_m \, \delta \left(t - m - \frac{n}{24} \right), \qquad (9.10)$$

where

$$a_m = \sum_{\substack{\mu \in \mathbb{Z}^n \\ |\mu|^2 = 24\,m+n}} \chi(\mu) \, P^k(\mu), \qquad (9.11)$$

is a τ-adapted distribution in the sense of Definition 9.1.

Proof. If P is a harmonic polynomial on $\mathbb{R}^n$, homogeneous of degree k, the Fourier transform of the function $x \mapsto P(x)\,U(|x|)$, where $x \mapsto U(|x|)$ is an arbitrary (except for questions of convergence) radial function, is the function $x \mapsto P(x)\,V(|x|)$, with

$$V(r) = 2\pi\, i^{-k} \int_0^\infty r^{\frac{2-n}{2}-k} \int_0^\infty U(\rho)\, \rho^{\frac{n}{2}+k} \, J_{\frac{n-2}{2}+k}(2\pi r\rho)\, d\rho, \qquad (9.12)$$

a formula which goes back to Hecke or Bochner. If one sets

$$v(s) = s^{\frac{n-2}{2}+k} \, V\left(\sqrt{2s}\right),$$
$$u(t) = t^{\frac{n-2}{2}+k} \, U\left(\sqrt{2t}\right), \qquad (9.13)$$

the formula becomes

$$v(s) = 2\pi\, i^{-k} \int_0^\infty u(t) \left(\frac{s}{t}\right)^{\frac{n-2}{4}+\frac{k}{2}} J_{\frac{n-2}{2}+k}(4\pi\sqrt{st}) : \qquad (9.14)$$

comparing the formula to (6.2), one may write it as

$$v = i^{-k}\, \sigma^\tau\, u \qquad (9.15)$$

with $\tau = \frac{n-2}{2} + k$. This interpretation of the Hankel transformation (6.2) is well known.

We now combine it with the consideration of the nth tensor power of the distribution $\mathfrak{d}_{\text{even}}$, *i.e.*, the distribution

$$\mathfrak{D}(x_1, \dots, x_n) = \mathfrak{d}_{\text{even}}(x_1) \dots \mathfrak{d}_{\text{even}}(x_n) : \qquad (9.16)$$

such a distribution on $\mathbb{R}^n$ is invariant under the multiplication by $e^{-\frac{i\pi n}{12}}\, e^{i\pi |x|^2}$: it is also invariant under the global Fourier transformation, up to some pth root of unity with $p = \frac{24}{(24,\,n)}$: this is so because of (3.19), together with the fact that each of the two transformations under consideration can be regarded as the product of n one-dimensional analogous transformations acting in each variable independently.

Let $d\sigma$ be the usual rotation-invariant measure on the unit sphere S^{n-1}, with total mass $\frac{2\pi^{\frac{n}{2}}}{\Gamma(\frac{n}{2})}$. On the sphere of radius $a > 0$, let us define the measure $d\sigma_a$ in such a way that the identity

$$\int_0^R \langle d\sigma_a, f \rangle \, da = \int_{|x| \leq R} f(x) \, dx \tag{9.17}$$

should hold for every summable function f, in other words

$$\langle d\sigma_a, f \rangle = a^{n-1} \int_{S^{n-1}} f(a\xi) \, d\sigma(\xi). \tag{9.18}$$

For every $k = 0, 1, \ldots$, let $(P_\alpha^k)_{\alpha \leq N_k}$ be a collection of harmonic polynomials, homogeneous of degree k, the restrictions Y_α^k of which to the unit sphere make up an orthonormal basis of the space of spherical harmonics of degree k. Then, the coefficients f_α^k of the decomposition

$$f(x) = \sum_{k=0}^\infty \sum_{\alpha \leq N_k} f_\alpha^k(|x|) \, P_\alpha^k(x) \tag{9.19}$$

of an arbitrary (say, continuous with compact support) function f on $\mathbb{R}^n$ are given by the equation

$$f_\alpha^k(a) = a^{-2k-n+1} \langle f, P_\alpha^k \, d\sigma_a \rangle, \qquad a > 0. \tag{9.20}$$

On the other hand, one can extend (9.17) to the case when f is replaced by a measure $\delta(x - x^0)$ with $x^0 \in \mathbb{R}^n \backslash \{0\}$, getting as a result

$$\langle \delta(x - x^0), d\sigma_a \rangle = \delta(a - |x^0|). \tag{9.21}$$

As a consequence, the measure $\delta(x - x^0)$ admits the decomposition on the right-hand side of (9.19) provided one sets

$$f_\alpha^k(a) = |x^0|^{-2k-n+1} \, P_\alpha^k(x^0) \, \delta(a - |x^0|). \tag{9.22}$$

The distribution $\mathfrak{D}$ thus decomposes as (9.19), with

$$f_\alpha^k(a) = \sum_{\mu = (\mu_1, \ldots, \mu_n) \in \mathbb{Z}^n} \chi(\mu) \left(\frac{|\mu|}{\sqrt{12}}\right)^{-2k-n+1} P_\alpha^k\left(\frac{\mu}{\sqrt{12}}\right) \delta\left(a - \frac{|\mu|}{\sqrt{12}}\right), \tag{9.23}$$

where $\chi(\mu)$ has been defined in (9.9). We now set $u(t) = t^{\frac{n-2}{2}+k} f_\alpha^k(\sqrt{2t})$ as in (9.13). Since $\delta\left(\sqrt{2t} - b\right) = b\,\delta(t - \frac{b^2}{2})$ for $b > 0$, we obtain

$$u(t) = 12^{-\frac{k}{2}} \times 2^{1-k-\frac{n}{2}} \sum_{\mu \in \mathbb{Z}^n} \chi(\mu) \, P_\alpha^k(\mu) \, \delta\left(t - \frac{|\mu|^2}{24}\right). \tag{9.24}$$

When $\chi(\mu) \neq 0$, one has $\mu_j^2 \equiv 1 \bmod 24$ for every j, so that $|\mu|^2 \equiv n \bmod 24$.

$\square$

In the case when P is the constant 2^{-n}, the modular form f associated under (9.2) to the τ-adapted distribution $\mathfrak{s}_P^0$ is

$$f(z) = 2^{-n} q^{\frac{n}{24}} \sum_{m \geq 0} a_m q^m$$

$$= 2^{-n} \sum_{(\mu_1, \ldots, \mu_n) \in \mathbb{Z}^n} \chi^{(12)}(\mu_1) \ldots \chi^{(12)}(\mu_n) \, q^{\frac{\mu_1^2 + \cdots + \mu_n^2}{24}}$$

$$= \left[\frac{1}{2} \sum_{\mu_1 \in \mathbb{Z}} \chi^{(12)}(\mu_1) q^{\frac{\mu_1^2}{24}} \right]^n$$

$$= (\eta(z))^n \tag{9.25}$$

according to (3.41). Hence, we are back to example (9.3), with $\tau = \frac{n-2}{2}$.

As in Section 5, we now set (*cf.* (7.13) for the extension of the representation $\mathcal{D}_{\tau+1}$ to distributions)

$$\mathfrak{s}_\tau^g = \mathcal{D}_{\tau+1}(g^{-1}) \mathfrak{s}_\tau, \qquad g \in SL(2, \mathbb{R}) : \tag{9.26}$$

that we denote this distribution as $\mathfrak{s}_\tau^g$ rather than $\mathfrak{s}_\tau^{\tilde{g}}$ for some $\tilde{g}$ lying above g in some group covering $SL(2, \mathbb{R})$ is due to the fact that, not having to worry about phase factors, we decided to consider $\mathcal{D}_{\tau+1}$ as a projective representation of $SL(2, \mathbb{R})$ rather than a genuine representation of, say, the universal cover of this group. Also note that, since we have dispensed with the quadratic transformation which would take the role previously taken by $\mathrm{Sq}_{\mathrm{even}}$ or $\mathrm{Sq}_{\mathrm{odd}}$, the distribution $\mathfrak{s}_\tau$ lives on the half-line, not the line.

We now show that, as g describes $\Gamma \backslash G$, the distributions $\mathfrak{s}_\tau^g$ constitute a set of (arithmetic) coherent states for the representation $\mathcal{D}_{\tau+1}$ in the sense described in the introduction.

Theorem 9.3. *Normalize the Haar measure dg on G as $dg = \frac{1}{2\pi} d\mu(z) \, d\theta$ if $x = x + iy$ and $g = \begin{pmatrix} y^{\frac{1}{2}} & xy^{-\frac{1}{2}} \\ 0 & y^{-\frac{1}{2}} \end{pmatrix} \begin{pmatrix} \cos\theta & -\sin\theta \\ \sin\theta & \cos\theta \end{pmatrix}, \ 0 \leq \theta < 2\pi$. Let $v \in C_{\tau+1}^\infty(\mathbb{R}^+)$, the space of C^∞ vectors of the representation $\mathcal{D}_{\tau+1}$. One has the identity*

$$\int_{\Gamma \backslash G} |(\mathfrak{s}_\tau^g \,|\, v)_{\tau+1}|^2 \, dg = \frac{(4\pi)^{\tau+1}}{\Gamma(\tau+1)} \, \| y^{\frac{\tau+1}{2}} f \|_{L^2(\Gamma \backslash \Pi)}^2 \, \| v \|_{\tau+1}^2, \tag{9.27}$$

where f is the modular form of weight $\tau + 1$ associated to $\mathfrak{s}_\tau$ by (9.2), so that the function $z \mapsto y^{\frac{\tau+1}{2}} |f(z)|$ is automorphic.

Proof. With $\mathfrak{s}_\tau$ as in (9.1), we first compute the integral

$$I(\tau) = \int_{\Gamma \backslash G} |(\mathfrak{s}_\tau^g \,|\, \psi_i^{\tau+1})_{\tau+1}|^2 \, dg, \tag{9.28}$$

where the distribution $\mathfrak{s}_\tau^g$ has been defined in (9.26) and, for our purpose, indeterminate phase factors depending only on g are harmless. One has

$$
I(\tau) = \int_{\Gamma \backslash G} |(\mathcal{D}_{\tau+1}(g^{-1})\,\mathfrak{s}_\tau \,|\, \psi_i^{\tau+1})_{\tau+1}|^2 \, dg
$$

$$
= \int_{\Gamma \backslash G} |(\mathfrak{s}_\tau \,|\, (\mathcal{D}_{\tau+1}(g)\,\psi_i^{\tau+1})_{\tau+1}|^2 \, dg = \int_{\Gamma \backslash G} |(\mathfrak{s}_\tau \,|\, \psi_{\frac{ai+b}{ci+d}}^{\tau+1})_{\tau+1}|^2 \, dg, \qquad (9.29)
$$

with $g = \left(\begin{smallmatrix} a & b \\ c & d \end{smallmatrix}\right)$, a consequence of (6.7). Next, with $z = g.i$, one has the identity

$$
(\mathfrak{s}_\tau \,|\, \psi_z^{\tau+1})_{\tau+1} = \frac{(4\pi)^{\frac{\tau+1}{2}}}{(\Gamma(\tau+1))^{\frac{1}{2}}} \left(\mathrm{Im}\left(-\frac{1}{z}\right)\right)^{\frac{\tau+1}{2}} \sum_{m \geq 0} \overline{a}_m \, e^{\frac{2i\pi(m+\kappa)}{\overline{z}}}
$$

$$
= \frac{(4\pi)^{\frac{\tau+1}{2}}}{(\Gamma(\tau+1))^{\frac{1}{2}}} \left(\mathrm{Im}\left(-\frac{1}{z}\right)\right)^{\frac{\tau+1}{2}} \overline{f\left(-\frac{1}{z}\right)} : \qquad (9.30)
$$

we have used (7.16), (9.1), (9.2). Hence,

$$
I(\tau) = \frac{(4\pi)^{\tau+1}}{\Gamma(\tau+1)} \int_{\Gamma \backslash \Pi} \left(\mathrm{Im}\left(-\frac{1}{z}\right)\right)^{\tau+1} \left|f\left(-\frac{1}{z}\right)\right|^2 \, d\mu(z)
$$

$$
= \frac{(4\pi)^{\tau+1}}{\Gamma(\tau+1)} \int_{\Gamma \backslash \Pi} y^{\tau+1} \, |f(z)|^2 \, d\mu(z) = \frac{(4\pi)^{\tau+1}}{\Gamma(\tau+1)} \, \| \, y^{\frac{\tau+1}{2}} \, f \, \|^2_{L^2(\Gamma \backslash \Pi)}. \qquad (9.31)
$$

Using (6.7) again, it is clear that one also has

$$
I(\tau) = \int_{\Gamma \backslash G} |(\mathfrak{s}_\tau^g \,|\, \psi_z^{\tau+1})_{\tau+1}|^2 \, dg \qquad (9.32)
$$

for every $z \in \Pi$. We now prove that, setting

$$
I_\tau(w, z) := \int_{\Gamma \backslash G} (\psi_w^{\tau+1} \,|\, \mathfrak{s}_\tau^g)_{\tau+1} \, (\mathfrak{s}_\tau^g \,|\, \psi_z^{\tau+1})_{\tau+1} \, dg, \qquad (9.33)
$$

one has

$$
I_\tau(w, z) = \frac{(4\pi)^{\tau+1}}{\Gamma(\tau+1)} \, \| \, y^{\frac{\tau+1}{2}} \, f \, \|^2_{L^2(\Gamma \backslash \Pi)} \, (\psi_w^{\tau+1} \,|\, \psi_z^{\tau+1})_{\tau+1} \qquad (9.34)
$$

for every pair (w, z) of points of Π. Note that, with $\phi_z^{\tau+1}$ as in Lemma 6.3,

$$
(\phi_w^{\tau+1} \,|\, \phi_z^{\tau+1}) = \int_0^\infty s^\tau \, e^{2i\pi s(\overline{z}^{-1} - w^{-1})} \, ds
$$

$$
= \frac{\Gamma(\tau+1)}{(2\pi)^{\tau+1}} \, e^{\frac{i\pi(\tau+1)}{2}} \left(\frac{w - \overline{z}}{\overline{z}\,w}\right)^{-\tau-1}, \qquad (9.35)
$$

so that, since

$$\psi_z^{\tau+1} = \frac{(4\pi)^{\frac{\tau+1}{2}}}{(\Gamma(\tau+1))^{\frac{1}{2}}} \left(\operatorname{Im} \left(-\frac{1}{z} \right) \right)^{\frac{\tau+1}{2}} \phi_z^{\tau+1}, \tag{9.36}$$

one has

$$(\psi_w^{\tau+1} \mid \psi_z^{\tau+1}) = 2^{\tau+1} e^{\frac{i\pi(\tau+1)}{2}} \left(\operatorname{Im} \left(-\frac{1}{w} \right) \right)^{\frac{\tau+1}{2}} \left(\operatorname{Im} \left(-\frac{1}{z} \right) \right)^{\frac{\tau+1}{2}} \left(\frac{w-\bar{z}}{\bar{z}w} \right)^{-\tau-1}. \tag{9.37}$$

Equation (9.34), which reduces to (9.32) when $w = z$, is correct in this case: to prove it in general, it thus suffices to prove that the product

$$\left(\operatorname{Im} \left(-\frac{1}{w} \right) \right)^{-\frac{\tau+1}{2}} \left(\operatorname{Im} \left(-\frac{1}{z} \right) \right)^{-\frac{\tau+1}{2}} I_\tau(w, z) \tag{9.38}$$

is a sesquiholomorphic function of (w, z) or, equivalently, that the integral

$$\int_{\Gamma \backslash G} (\phi_w^{\tau+1} \mid \mathfrak{s}_\tau^g)_{\tau+1} (\mathfrak{s}_\tau^g \mid \phi_z^{\tau+1})_{\tau+1} \, dg \tag{9.39}$$

is a sesquiholomorphic function of (w, z). Since, from (6.9), $\phi_z^{\tau+1}$ is an antiholomorphic function of z, it follows from (6.10) that so is, for fixed g, the function $\mathcal{D}_{\tau+1}(g) \phi_z^{\tau+1}$. This proves (9.34).

Equation (9.27) follows in the case when v lies in the linear space generated by the functions $\phi_z^{\tau+1}$, and Theorem 9.3 is obtained by completion: the assumption that v lies in the space of C^∞ vectors of the representation $\mathcal{D}_{\tau+1}$ gives the scalar product $(\mathfrak{s}_\tau^g \mid v)$ a meaning for every $g \in G$. $\qquad \square$

As a preparation towards the next section, we transform in a classical way [4, p. 70] the function

$$L(\bar{f} \otimes f, s) = \sum_{n \geq 0} |a_n|^2 (n+\kappa)^{-s-\tau} : \tag{9.40}$$

it is a convolution L-function, and the notation agrees, in the case when $\tau+1$ is an even integer and $\kappa = 1$, with that in [13, p. 249]. The function

$$\phi(z) = y^{\tau+1} |f(z)|^2$$
$$= y^{\tau+1} \sum_{m,n \geq 0} \bar{a}_m a_n \exp 2i\pi \left(-(m+\kappa) \bar{z} + (n+\kappa) z \right) \tag{9.41}$$

is automorphic with respect to the full modular group, and rapidly decreasing at infinity in the fundamental domain. Applying the Rankin-Selberg transformation, recalled in Proposition 8.1, one obtains if $\operatorname{Re} s > 1$ the equation

$$\int_{\mathcal{D}} y^s \phi(z) \, d\mu(z) = \int_{\Gamma \backslash \Pi} E(z, s) \phi(z) \, d\mu(z). \tag{9.42}$$

Now, from (9.41), one has

$$\int_{-\frac{1}{2}}^{\frac{1}{2}} \phi(x + iy)\, dx = y^{\tau+1} \sum_{n \geq 0} |a_n|^2\, e^{-4\pi\,(n+\kappa)y} , \tag{9.43}$$

so that the left-hand side of (9.42) reduces to

$$\int_{0}^{\infty} y^{s+\tau-1} \sum_{n \geq 0} |a_n|^2\, e^{-4\pi\,(n+\kappa)y}\, dy = \frac{\Gamma(s+\tau)}{(4\pi)^{s+\tau}} \sum_{n \geq 0} |a_n|^2\, (n+\kappa)^{-s-\tau} . \tag{9.44}$$

Note, incidentally, that the convergence of the series on the right-hand side is obtained for Re $s > 1$, whereas an application of (9.8) would only ascertain it for Re $s > 2$. The final equation

$$\int_{\Gamma\backslash\Pi} E(z, s)\, \phi(z)\, d\mu(z) = \frac{\Gamma(s+\tau)}{(4\pi)^{s+\tau}} \sum_{n \geq 0} \frac{|a_n|^2}{(n+\kappa)^{s+\tau}} \tag{9.45}$$

shows that the function

$$L^*(\bar{f} \otimes f,\, s) = \zeta^*(2s) \times \frac{\Gamma(s+\tau)}{(4\pi)^{s+\tau}} \sum_{n \geq 0} \frac{|a_n|^2}{(n+\kappa)^{s+\tau}} \tag{9.46}$$

extends as a meromorphic function in the entire complex plane, with simple poles only at $s = 0$ and $s = 1$, and is invariant under the change $s \mapsto 1 - s$. All this is quite classical, except for the presence of κ.

We would like to know more than we do about the zeros of $L(\bar{f} \otimes f,\, s)$ on the critical line, at least in some instances.

Proposition 9.4. *In the case when $f = f^{(12)}$, as made explicit in (3.42), the zeros on the line Re $s = \frac{1}{2}$ of the function $L(\bar{f} \otimes f,\, s)$ are just the solutions of the (elementary) equations $2^{1-2s} = 1$ and $3^{1-2s} = 1$; in the case when $f = f^{(4)}$ as in (3.44), they make up the set of zeros of the first of these two equations. In the case when $f(z) = q \prod_{n\geq 1}(1 - q^n)^{24} = (2\pi)^{-12}\, \Delta(z)$, the Ramanujan delta function, the critical zeros of the function $L(\bar{f} \otimes f,\, s)$ include those of the Riemann zeta function.*

Proof. In the first case, one has $\tau = -\frac{1}{2}$, $\kappa = \frac{1}{24}$ and $a_m = \pm 1$ if $m = \frac{\ell(3\ell-1)}{2}$ for some $\ell \in \mathbb{Z}$, while $m = 0$ otherwise: since $\frac{\ell(3\ell-1)}{2} + \frac{1}{24} = \frac{(6\ell-1)^2}{24}$, one obtains

$$L(\bar{f} \otimes f,\, s) = \sum_{\ell \in \mathbb{Z}} \left[\frac{(6\ell - 1)^2}{24} \right]^{\frac{1}{2}-s} = 24^{s-\frac{1}{2}} \sum_{\substack{n \geq 1 \\ (n,6)=1}} n^{1-2s}$$

$$= 24^{s-\frac{1}{2}} \left[1 - 2^{1-2s} - 3^{1-2s} + 6^{1-2s} \right] \zeta(2s - 1)$$

$$= 24^{s-\frac{1}{2}} \left(1 - 2^{1-2s} \right) \left(1 - 3^{1-2s} \right) \zeta(2s - 1) : \tag{9.47}$$

the residue of this function at $s = 1$ is $(\frac{2}{3})^{\frac{1}{2}}$. In the second case, one has $\tau = \frac{1}{2}$, $\kappa = \frac{1}{8}$ and $a_m = \pm(n + \frac{1}{2})$ if $m = \frac{n(n+1)}{2}$ with $n \geq 0$, while $a_m = 0$ otherwise: then, $\frac{n(n+1)}{2} + \frac{1}{8} = \frac{(2n+1)^2}{8}$, so that

$$L(\bar{f} \otimes f, s) = \sum_{n \geq 0} \left(n + \frac{1}{2}\right)^2 \left[\frac{(2n+1)^2}{8}\right]^{-s-\frac{1}{2}}$$

$$= 2^{-\frac{1}{2}+3s} \sum_{n \geq 0} (2n+1)^{1-2s}$$

$$= 2^{-\frac{1}{2}+3s} \left(1 - 2^{1-2s}\right) \zeta(2s-1)$$

$$= 2^{\frac{1}{2}+s} \left(2^{2s-1} - 1\right) \zeta(2s-1) : \qquad (9.48)$$

the residue of this function at $s = 1$ is $2^{\frac{1}{2}}$.

In the third case, one has $\tau = 11$, $\kappa = 1$, and the result can be found in the last chapter of Iwaniec's book [13], where it is ascribed to Shimura [23]. In (9.2), we chose to write $f(z) = q \sum_{m \geq 0} a_m\, q^m$, and a notation closer to the one in use in [13] would be

$$f(z) = \sum_{n \geq 1} c(n)\, n^{\frac{7}{2}}\, q^n. \qquad (9.49)$$

From a relation between the Fourier coefficients of f provided by the use of the theory of Hecke operators, it is proved in [13, p. 250] that

$$L(f \otimes f, s) = \zeta(s)\, Z_f(s) \qquad (9.50)$$

with

$$Z_f(s) = \sum_{n \geq 1} c(n^2)\, n^{-s} : \qquad (9.51)$$

also, it is shown (*loc. cit.*, p. 247) that the function Z_f is holomorphic in a neighbourhood of the closed half-plane Re $s \geq \frac{1}{2}$. Consequently, $L(f \otimes f, s)$ vanishes at all zeros s of the zeta function such that Re $s \geq \frac{1}{2}$. A similar result continues to hold whenever f is a Hecke cusp-form of the usual kind (*i.e.*, of even integral weight) for the full modular group. $\qquad \square$

Proposition 9.5. *One may rewrite* (9.27) *as*

$$\int_{\Gamma \backslash G} |(\mathfrak{s}_\tau^g \,|\, u)_{\tau+1}|^2 \, dg = \frac{\pi}{3} \operatorname{Res}_{s=1} \left(L(\bar{f} \otimes f, s)\right) \|u\|_{\tau+1}^2; \qquad (9.52)$$

in particular, when $\tau = -\frac{1}{2}$ *(resp.* $\tau = \frac{1}{2}$*), the coefficient on the right-hand side is* $\pi \sqrt{\frac{2}{27}}$ *(resp.* $\frac{\pi\sqrt{2}}{3}$ *).*

Proof. It is an immediate consequence of (9.45), (9.46) and of the classical Fourier expansion of Eisenstein series [12, p. 68] recalled in (10.50) below that

$$\text{Res}_{s=1}\left(L(\bar{f} \otimes f,\, s)\right) = \frac{3}{\pi}\frac{(4\pi)^{\tau+1}}{\Gamma(\tau+1)}\int_{\Gamma\backslash\Pi} y^{\tau+1}\,|f(z)|^2\,d\mu(z)$$

$$= \frac{3}{\pi}\frac{(4\pi)^{\tau+1}}{\Gamma(\tau+1)}\,\|\,y^{\frac{\tau+1}{2}}\,f\,\|^2_{L^2(\Gamma\backslash\Pi)}\;: \qquad (9.53)$$

recall that the function $z \mapsto y^{\frac{\tau+1}{2}}\,|f(z)|$ is automorphic. This equation can be found in [13, p. 246]: whether $\tau+1$ is an integer or not does not change anything at this point.   □

Remark 9.1. Equations (4.1) and (4.2) follow: the apparent discrepancy by a factor $2^{\frac{3}{2}}$ in the first case, $2^{\frac{1}{2}}$ in the second one, come from the fact that, with the notation in Section 3, $\mathfrak{d}^{(12)}$ (*resp.* $\mathfrak{d}^{(4)}$) is the image of the distribution $\mathfrak{S}^{(12)}$ (*resp.* $\mathfrak{S}^{(4)}$) there under the map $2^{\frac{3}{4}}\,\text{Sq}_{\text{even}}$ (*resp.* $2^{\frac{1}{4}}\,\text{Sq}_{\text{odd}}$).

A question. Looking back at Theorem 7.7, we observe that it is the τ-dependent operator $\Gamma(\tau+\frac{1}{2}+i\pi\mathcal{E})$ that enters the formula. Simultaneously using Corollary 5.2, one might wish for a formula in which, for some real a, the operator $\zeta(a - 2i\pi\mathcal{E})$ would play an analogous role, possibly preparing the way for some new interpretation of the zeros of the zeta function on any *given* line. However, as will be seen in the next section, it is the function $L(\bar{f} \otimes f,\, s)$, taken on the line $\text{Re } s = \frac{1}{2}$, that enters the spectral density relative to the automorphic function under study. Shimura's result recalled in Proposition 9.4 shows that this function is divisible by $\zeta(s)$ on the critical line, in the case when $\tau+1$ is an even integer and f is a Hecke cusp-form for the full modular group. Does this property, changing the critical line for another one, extend, in the case when $f = \eta^{2\tau+2}$ is a power of Dedekind's eta-function, to exponents distinct from 11 ? In the case when $\tau = \frac{n-2}{2}$ with $n = 1, 2, \ldots$, it follows from (9.11) (with $P^k = 2^{-n}$) that

$$L(\bar{f} \otimes f,\, s) = 2^{-2n} \times 24^{s+\frac{n-2}{2}} \sum_{\substack{\mu,\nu\in\mathbb{Z}^n \\ |\mu|^2=|\nu|^2}} \frac{\chi(\mu)\,\chi(\nu)}{|\mu|^{2s+n-2}}\;: \qquad (9.54)$$

setting $(\mu,\, \nu) = g.c.d.\,(\mu_1,\ldots,\mu_n,\nu_1,\ldots,\nu_n)$, one can write

$$L(\bar{f} \otimes f,\, s) = 24^{s+\frac{n-2}{2}}\,(1 - 2^{2-n-2s})\,(1 - 3^{2-n-2s})\,\zeta(2s+n-2)\,F_{\frac{n-2}{2}}(s) \quad (9.55)$$

with

$$F_{\frac{n-2}{2}}(s) = 2^{-2n} \sum_{\substack{\mu,\nu\in\mathbb{Z}^n \\ |\mu|^2=|\nu|^2 \\ (\mu,\nu)=1}} \frac{\chi(\mu)\,\chi(\nu)}{|\mu|^{2s+n-2}}\,. \qquad (9.56)$$

Of course, the fact that the factor $\zeta(2s + n - 2)$ appears "naturally" here does not imply that if one substitutes τ for $\frac{n-2}{2}$, the function $L(\bar{f} \otimes f, s)$ will be "divisible" by $\zeta(2s + 2\tau)$ in a sufficient domain, especially when $\tau < -\frac{1}{2}$: only think of the continuation of the Eisenstein series $E(z, s)$ as opposed to that of $E^*(z, s)$. Besides, this zeta factor is the "interesting" one when $n = 1$, not when $n = 11$. Our sole point here is that a shift (by $\tau + \frac{1}{2}$) in the argument of the Gamma function that occurs, in a spectral-theoretic role, in Theorem 7.7, is possible, only replacing the usual analysis on the line by the radial part of a fractional-dimensional theory; it would be nice if something similar could be done with the zeta function as it occurs in Corollary 5.2, but nothing is clear to us yet in this direction.

10 Radial horocyclic calculus and arithmetic

The present section is similar to Section 5: only, we trade the Weyl calculus for the τ-pseudodifferential (horocyclic) calculus. Using a calculus based on the use of Π as a phase space would not make the analysis that follows possible. The reason is that radial functions in the plane, at the same time generalized eigenfunctions of the Euler operator, are of course very simple, while the radial eigenfunctions of Δ, on Π, are Legendre functions of the function $z \mapsto \cosh d(i, z)$. Trying to analyze series of terms involving Legendre functions, the arguments of which would be (z-dependent) fractional-linear functions of a pair (m, n) of integers, would be a hopeless task.

Recall from (9.26) the definition of the distributions $\mathfrak{s}_\tau^g$. In accordance with the definition of $P_{v, u}$ just before (7.20), we set

$$P_{\mathfrak{s}_\tau^g, \mathfrak{s}_\tau^g}\, w = (\mathfrak{s}_\tau^g \,|\, w)_{\tau+1}\, \mathfrak{s}_\tau^g, \tag{10.1}$$

a well-defined operator from $C_{\tau+1}^\infty(\mathbb{R}^+)$ to $C_{\tau+1}^{-\infty}(\mathbb{R}^+)$ (again, $cf.$ (7.16) for a proper understanding of the scalar product in the last equation or the following one) invariant under the multiplication of $\mathfrak{s}_\tau^g$ by any phase factor. One has in particular

$$P_{\mathfrak{s}_\tau, \mathfrak{s}_\tau} = \sum_{m, n \geq 0} \bar{a}_m\, a_n\, \langle \delta_{m+\kappa}, \cdot \rangle_{\tau+1}\, \delta_{n+\kappa}. \tag{10.2}$$

We are now ready to come to the "arithmetic" calculations. At least temporarily, we shall use the soft horocyclic calculus, which is most suitable in view of Theorem 7.6. The τ-dependent generalization of the function $z \mapsto (\mathcal{A}h)_0(z) = (\partial_{\text{even}}^g \,|\, \text{Op}(h)\, \partial_{\text{even}}^g)$ as it has been defined in (5.3) in connection to the Weyl calculus is now defined (with $z = g.i$, still under the assumption that h is a radial function) as

$$(\mathcal{A}_\tau h)(z) = \text{Tr}\,(\text{Op}_{\text{soft}}^\tau(h)\, P_{\mathfrak{s}_\tau^g, \mathfrak{s}_\tau^g}) = \langle\, h, W^\tau(\mathfrak{s}_\tau^g, \mathfrak{s}_\tau^g)\, \rangle$$
$$= \langle\, h, W^\tau(\mathfrak{s}_\tau, \mathfrak{s}_\tau) \circ g\, \rangle. \tag{10.3}$$

Theorem 10.1. *The function* α_k *introduced at the end of Theorem 8.2 is given by the equation* (8.55)

$$\alpha_k(s) = \frac{1}{2} \frac{\Gamma(s - \frac{1}{2}) \Gamma(\frac{1}{2} - s)}{\zeta^*(2 - 2s)} k^{s - \frac{1}{2}} \sigma_{1-2s}(k) \,. \tag{10.4}$$

On the other hand, let $h \in \mathcal{S}(\mathbb{R}^2)$ *be a radial and* $\mathcal{G}$-*antiinvariant function; let* $\tau > -1$ *be given. Assume that*

$$h = i\pi \mathcal{E} \, h_1 \tag{10.5}$$

for some radial $\mathcal{G}$-*invariant function* $h_1 \in \mathcal{S}(\mathbb{R}^2)$. *Let* $\psi(s)$ *be defined for* Re $s < 1$, *as in* (5.11), *by the equation, in which* $h(q, p) = H(q^2 + p^2)$,

$$\psi(s) = \frac{1}{2\pi} \int_0^\infty r^{-s} \, H(r^2) \, dr \,. \tag{10.6}$$

The spectral decomposition of the automorphic function $\mathcal{A}_\tau \, h$ *introduced in* (10.3) *is given by the formula*

$$(\mathcal{A}_\tau \, h)(z)$$
$$= \frac{2\pi^{\frac{1}{2}}}{i} \int_{\frac{1}{2} - i\infty}^{\frac{1}{2} + i\infty} \psi(1 - 2s) \, \frac{\Gamma(s - \frac{1}{2}) \Gamma(\frac{3}{2} - s) \Gamma(s + \tau)}{2^{s+\tau} \, \Gamma(s) \Gamma(\tau + \frac{3}{2})} \, \frac{L(\overline{f} \otimes f, s)}{\zeta^*(2 - 2s)} \, E^*(z, s) \, ds$$
$$- 2^{-\tau - 1} \pi^{\frac{1}{2}} \frac{\Gamma(\tau + 1)}{\Gamma(\tau + \frac{3}{2})} \, h(0) \, \mathrm{Res}_{s=1} \left(L(\overline{f} \otimes f, s) \right) : \quad (10.7)$$

in particular, no cusp-forms enter it. The right-hand side is still meaningful if one dispenses with the assumption (10.5).

Proof. Choosing $g = \begin{pmatrix} y^{\frac{1}{2}} & xy^{-\frac{1}{2}} \\ 0 & y^{-\frac{1}{2}} \end{pmatrix}$, so that $z = g.i$, and $g \left({q \atop p} \right) = y^{-\frac{1}{2}} \left({yq + xp \atop p} \right)$, one writes, starting from (10.3),

$$(\mathcal{A}_\tau \, h)(z) = \left\langle h, \, W^\tau(\mathfrak{s}_\tau, \mathfrak{s}_\tau)(q, p) \mapsto \left(\frac{yq + xp}{y^{\frac{1}{2}}}, \frac{p}{y^{\frac{1}{2}}} \right) \right\rangle, \tag{10.8}$$

an expression which we immediately transform to

$$(\mathcal{A}_\tau \, h)(z) = \left\langle h, \, W^\tau(\mathfrak{s}_\tau, \mathfrak{s}_\tau)(q, p) \mapsto \left(-\frac{p}{y^{\frac{1}{2}}}, \frac{yq + xp}{y^{\frac{1}{2}}} \right) \right\rangle : \tag{10.9}$$

that this does not change the result is again a consequence of the covariance of the horocyclic calculus and of the invariance, up to some constant unitary factor, of $\mathfrak{s}_\tau$ under $\mathcal{D}_{\tau+1} \left(\left({0 \atop -1} {1 \atop 0} \right) \right)$.

As a consequence of Theorem 7.6, one has

$$
W^\tau(\delta_{m+\kappa}, \delta_{n+\kappa})\left(-\frac{p}{y^{\frac{1}{2}}}, \frac{yq+xp}{y^{\frac{1}{2}}}\right) = \exp\left(2i\pi(n-m)\frac{yq+xp}{p}\right)
$$
$$
\times \left(\frac{p^2}{y}\right)^\tau \left[\frac{p^4}{y^2} - \frac{(m-n)^2}{4}\right]\left[\frac{p^4}{y^2} + (m+n+2\kappa)\frac{p^2}{y} + \frac{(m-n)^2}{4}\right]^{-\tau-\frac{3}{2}}
$$
$$
(10.10)
$$

if $m \neq n$, while

$$
W^\tau(\delta_{n+\kappa}, \delta_{n+\kappa})\left(-\frac{p}{y^{\frac{1}{2}}}, \frac{yq+xp}{y^{\frac{1}{2}}}\right) = \frac{|p|}{y^{\frac{1}{2}}}\left(\frac{p^2}{y} + 2n + 2\kappa\right)^{-\tau-\frac{3}{2}}
$$
$$
- y^{\frac{1}{2}} \delta(p) \int_{-\infty}^\infty e^{-4i\pi ty^{\frac{1}{2}}q} |t| (t^2 + 2n + 2\kappa)^{-\tau-\frac{3}{2}} dt. \quad (10.11)
$$

Let us immediately observe that the second term of the last sum is the negative of the $\mathcal{G}$-transform of the first one, hence yields the same result when tested against a $\mathcal{G}$-antiinvariant symbol h.

We expand $(\mathcal{A}_\tau h)(z)$ as a series $\sum_{k \in \mathbb{Z}} F_k(z)$, in which the kth term is obtained by grouping all terms from the sum in (10.2) such that $m - n = k$. From (10.9), (10.10) and (10.11), we obtain

$$
F_k(z) = \sum_{\substack{m,n \geq 0 \\ m-n=k}} \int_{\mathbb{R}^2} h(q, p) \exp\left(-2i\pi k\frac{yq+xp}{p}\right)\left(\frac{p^2}{y}\right)^\tau \left[\frac{p^4}{y^2} - \frac{k^2}{4}\right]
$$
$$
\left[\frac{p^4}{y^2} + (m+n+2\kappa)\frac{p^2}{y} + \frac{k^2}{4}\right]^{-\tau-\frac{3}{2}} dq\, dp \quad (10.12)
$$

if $k \neq 0$, while

$$
F_0(z) = 2\, y^{-\frac{1}{2}} \sum_{n \geq 0} |a_n|^2 \int_{\mathbb{R}^2} h(q, p)\, |p| \left(\frac{p^2}{y} + 2n + 2\kappa\right)^{-\tau-\frac{3}{2}} dq\, dp. \quad (10.13)
$$

Let us write
$$
F_k(z) = F_k^{(1)}(z) - F_k^{(2)}(z), \quad (10.14)
$$

where the two terms of the decomposition are those corresponding to the decomposition of the sum $\frac{p^4}{y^2} - \frac{k^2}{4}$.

Setting $h(q, p) = H(q^2 + p^2)$, we use the decomposition (5.15) of f into homogeneous functions, here recalled for convenience:

$$
H(\rho) = \frac{2}{i} \int_{\sigma-i\infty}^{\sigma+i\infty} \psi(1 - 2s)\, \rho^{-s}\, ds, \qquad \sigma > 0. \quad (10.15)
$$

We also set $h_1(q, p) = H_1(q^2 + p^2)$: with

$$\psi(s) = \frac{1}{4\pi} \int_0^\infty \rho^{-\frac{s+1}{2}} H(\rho)\, d\rho, \quad \psi_1(s) = \frac{1}{4\pi} \int_0^\infty \rho^{-\frac{s+1}{2}} H_1(\rho)\, d\rho, \qquad (10.16)$$

as in (5.11), the equation $H = \left(\frac{1}{2} + \rho \frac{d}{d\rho}\right) H_1$ yields

$$\psi(s) = \frac{s}{2}\, \psi_1(s). \tag{10.17}$$

Since h is assumed to be $\mathcal{G}$ -antiinvariant, it follows from (5.29) that

$$(2\pi)^{1-s}\, \frac{\psi(2s-1)}{\Gamma(1-s)} = -(2\pi)^s\, \frac{\psi(1-2s)}{\Gamma(s)}. \tag{10.18}$$

Equation (10.15) leads to writing

$$F_0(z) = \frac{2}{i} \int_{\sigma - i\infty}^{\sigma + i\infty} \psi(1 - 2s)\, G_0(z; s)\, ds \tag{10.19}$$

and, if $k \neq 0$,

$$F_k^{(1)}(z) = \frac{2}{i} \int_{\sigma_1 - i\infty}^{\sigma_1 + i\infty} \psi(1 - 2s)\, G_k^{(1)}(z; s)\, ds,$$

$$F_k^{(2)}(z) = \frac{2}{i} \int_{\sigma_2 - i\infty}^{\sigma_2 + i\infty} \psi(1 - 2s)\, G_k^{(2)}(z; s)\, ds, \tag{10.20}$$

all three functions $G_0(z; s)$, $G_k^{(1)}(z; s)$, $G_k^{(2)}(z; s)$ having been obtained from $F_0(z)$, $F_k^{(1)}(z)$, $F_k^{(2)}(z)$ by substituting $(q^2 + p^2)^{-s}$ for $h(q, p)$: note that σ , σ_1 , σ_2 can be chosen independently as soon as convergence is ensured, subject to the restriction that they must be > 0 .

   As is always the case when dealing with non-holomorphic modular forms, we shall have to use the following, taken from [12, p. 66] or [16, p.401]:

Lemma 10.2. *If* $\mathrm{Re}\, s > \frac{1}{2}$, $\alpha > 0$ *and* $\beta > 0$, *one has*

$$\int_{-\infty}^\infty (\theta^2 + \alpha^2)^{-s}\, e^{2i\pi\beta\theta}\, d\theta = \frac{2\pi^s}{\Gamma(s)} \left(\frac{\beta}{\alpha}\right)^{s-\frac{1}{2}} K_{-\frac{1}{2}+s}(2\pi\alpha\beta); \tag{10.21}$$

if $\mathrm{Re}\, s > \frac{1}{2}$, $\alpha > 0$ *and* $\beta = 0$, *the value of the same integral is* $\pi^{\frac{1}{2}} \frac{\Gamma(s-\frac{1}{2})}{\Gamma(s)}\, \alpha^{1-2s}$.

   We study $G_0(z; s)$ first. After one has applied the lemma, it reduces to

$$G_0(z; s) = 2\,\pi^{\frac{1}{2}}\, \frac{\Gamma(s - \frac{1}{2})}{\Gamma(s)}\, y^{-\frac{1}{2}} \sum_{n \geq 0} |a_n|^2$$

$$\times \int_{-\infty}^\infty |p|^{2-2s} \left(\frac{p^2}{y} + 2n + 2\kappa\right)^{-\tau - \frac{3}{2}} dp. \tag{10.22}$$

The last integral, convergent if $\mathrm{Re}\ s < \frac{3}{2}$ and $\tau + \mathrm{Re}\ s > 0$, can be written as

$$y^{\frac{3}{2}-s} \int_{-\infty}^{\infty} |p|^{2-2s} (p^2 + 2n + 2\kappa)^{-\tau-\frac{3}{2}}\, dp$$

$$= y^{\frac{3}{2}-s} \frac{\Gamma(\frac{3}{2}-s)\,\Gamma(\tau+s)}{\Gamma(\tau+\frac{3}{2})} (2n+2\kappa)^{-\tau-s}, \quad (10.23)$$

so that

$$G_0(z;\, s) = 2\,\pi^{\frac{1}{2}} \frac{\Gamma(s-\frac{1}{2})\,\Gamma(\frac{3}{2}-s)\,\Gamma(\tau+s)}{\Gamma(s)\,\Gamma(\tau+\frac{3}{2})}\, y^{1-s}$$

$$\times\, 2^{-\tau-s} \sum_{n\geq 0} |a_n|^2\, (n+\kappa)^{-\tau-s} : \quad (10.24)$$

in view of (9.44), the last series converges provided that $\mathrm{Re}\ s > 1$: the only other constraint is $\mathrm{Re}\ s < \frac{3}{2}$, since the condition $\tau + \mathrm{Re}\ s > 0$ automatically holds when $\mathrm{Re}\ s > 1$ in view of the standing assumption $\tau > -1$.

It then follows from (10.19) and (9.40) that

$$F_0(z) = \frac{4}{i} \int_{\sigma-i\infty}^{\sigma+i\infty} \psi(1-2s)\, 2^{-\tau-s}\, y^{1-s}$$

$$\pi^{\frac{1}{2}} \frac{\Gamma(s-\frac{1}{2})\,\Gamma(\frac{3}{2}-s)\,\Gamma(\tau+s)}{\Gamma(s)\,\Gamma(\tau+\frac{3}{2})}\, L(\overline{f} \otimes f,\, s)\, ds \quad (10.25)$$

or, using (9.46),

$$F_0(z) = \frac{2^{\frac{3}{2}}}{i} \int_{\sigma-i\infty}^{\sigma+i\infty} \psi(1-2s)\, \frac{(2\pi)^{s+\tau+\frac{1}{2}}}{\zeta^*(2s)}\, \frac{\Gamma(s-\frac{1}{2})\,\Gamma(\frac{3}{2}-s)}{\Gamma(s)\,\Gamma(\tau+\frac{3}{2})}\, y^{1-s}\, L^*(\overline{f} \otimes f,\, s)\, ds.$$

$$(10.26)$$

One moves now the line of integration from $\sigma + i\mathbb{R}$ to $\frac{1}{2} + i\mathbb{R}$: first, according to (9.46), the only pole of $L^*(\overline{f} \otimes f,\, s)$ crossed along the way is $s = 1$: starting from (10.25) and using also (5.12) and the fact that h is $\mathcal{G}$-antiinvariant, one sees that this pole contributes the extra term

$$\mathrm{Extra} = -2^{-\tau-1}\, \pi^{\frac{1}{2}}\, \frac{\Gamma(\tau+1)}{\Gamma(\tau+\frac{3}{2})}\, h(0)\, \mathrm{Res}_{s=1} (L(\overline{f} \otimes f,\, s)). \quad (10.27)$$

In the main term, the integration takes place on the line $\frac{1}{2} + i\mathbb{R}$: changing s to $1-s$ and using (10.18) together, one may also write the integral term as

$$-\frac{2^{\frac{3}{2}}}{i} \int_{\frac{1}{2}-i\infty}^{\frac{1}{2}+i\infty} \psi(1-2s)\, \frac{(2\pi)^{s+\tau+\frac{1}{2}}}{\zeta^*(2-2s)}\, \frac{\Gamma(\frac{1}{2}+s)\,\Gamma(\frac{1}{2}-s)}{\Gamma(s)\,\Gamma(\tau+\frac{3}{2})}\, y^s\, L^*(\overline{f} \otimes f,\, 1-s)\, ds :$$

$$(10.28)$$

we use the equation $L^*(\overline{f} \otimes f, 1-s) = L^*(\overline{f} \otimes f, s)$ together with the fact that $\Gamma(\frac{1}{2}+s)\Gamma(\frac{1}{2}-s) = -\Gamma(s-\frac{1}{2})\Gamma(\frac{3}{2}-s)$, then add (10.26), taken with $\sigma = \frac{1}{2}$, and (10.28): the final result is

$$F_0(z) = \frac{2^{\frac{1}{2}}}{i} \int_{\frac{1}{2}-i\infty}^{\frac{1}{2}+i\infty} \psi(1-2s)\,(2\pi)^{s+\tau+\frac{1}{2}} \, \frac{\Gamma(s-\frac{1}{2})\,\Gamma(\frac{3}{2}-s)}{\Gamma(s)\,\Gamma(\tau+\frac{3}{2})}$$

$$\left[\frac{y^{1-s}}{\zeta^*(2s)} + \frac{y^s}{\zeta^*(2-2s)}\right] L^*(f \otimes f, s)\, ds$$

$$- 2^{-\tau-1}\,\pi^{\frac{1}{2}} \, \frac{\Gamma(\tau+1)}{\Gamma(\tau+\frac{3}{2})}\, h(0)\, \mathrm{Res}_{s=1}\left(L(\overline{f} \otimes f, s)\right). \quad (10.29)$$

Note that $\frac{y^{1-s}}{\zeta^*(2s)} + \frac{y^s}{\zeta^*(2-2s)}$ is the "constant" term (*i.e.*, the term independent of x) in the Fourier expansion of $\frac{E(z,s)}{\zeta^*(2-2s)} = \frac{E(z,1-s)}{\zeta^*(2s)}$.

We now study the terms $G_k^{(1)}$ and $G_k^{(2)}$, assuming $k > 0$ since the case when $k < 0$ is fully similar. Starting from (10.12) with $(q^2+p^2)^{-s}$ substituted for $h(q,p)$, making the result of the dq-integration explicit with the help of Lemma 10.1 and performing the change $p \mapsto y^{\frac{1}{2}}p$ in the result, we obtain

$$G_k^{(1)}(z;s) = \frac{2\pi^s}{\Gamma(s)}\, k^{s-\frac{1}{2}}\, y^{\frac{1}{2}}\, K_{s-\frac{1}{2}}(2\pi ky)\, e^{-2i\pi kx}$$

$$\times \sum_{\substack{m,n \geq 0 \\ m-n=k}} \overline{a}_m\, a_n \int_{-\infty}^{\infty} |p|^{2\tau-2s+5} \left[p^4 + (m+n+2\kappa)\,p^2 + \frac{k^2}{4}\right]^{-\tau-\frac{3}{2}} dp,$$

$$(10.30)$$

where the integral converges if $1 < \mathrm{Re}\; s < 2$; similarly,

$$G_k^{(2)}(z;s) = \frac{k^2}{4} \times \frac{2\pi^s}{\Gamma(s)}\, k^{s-\frac{1}{2}}\, y^{\frac{1}{2}}\, K_{s-\frac{1}{2}}(2\pi ky)\, e^{-2i\pi kx}$$

$$\times \sum_{\substack{m,n \geq 0 \\ m-n=k}} \overline{a}_m\, a_n \int_{-\infty}^{\infty} |p|^{2\tau-2s+1} \left[p^4 + (m+n+2\kappa)\,p^2 + \frac{k^2}{4}\right]^{-\tau-\frac{3}{2}} dp,$$

$$(10.31)$$

where the integral converges if $-1 < \mathrm{Re}\; s < 0$.

Using the notation (8.4) and changing p to t^2 when $p > 0$, we obtain

$$G_k^{(1)}(z;s) = \frac{2\pi^s}{\Gamma(s)}\, k^{s-\frac{1}{2}}\, h_{s,-k}(z) \sum_{\substack{m,n \geq 0 \\ m-n=k}} \overline{a}_m\, a_n\, C_{\tau+1}(m,n;s-1),$$

$$G_k^{(2)}(z;s) = \frac{k^2}{4} \times \frac{2\pi^s}{\Gamma(s)}\, k^{s-\frac{1}{2}}\, h_{s,-k}(z) \sum_{\substack{m,n \geq 0 \\ m-n=k}} \overline{a}_m\, a_n\, C_{\tau+1}(m,n;s+1), \quad (10.32)$$

with

$$C_\tau(m,\, n;\, s) = \int_0^\infty t^{\tau-s}\, [t^2 + At + B]^{-\tau-\frac{1}{2}}\, dt\,, \tag{10.33}$$

where we have set $A = m + n + 2\kappa$ and $B = \frac{k^2}{4}$. We shall transform this expression, starting from

$$[t^2 + At + B]^{-\tau-\frac{1}{2}} = \frac{(2\pi)^{\tau+\frac{1}{2}}}{\Gamma(\tau+\frac{1}{2})} \int_0^\infty e^{-2\pi v(t^2 + At + B)}\, v^{\tau-\frac{1}{2}}\, dv\,, \tag{10.34}$$

so that

$$
\begin{aligned}
C_\tau(m,\, n;\, s) &= \int_0^\infty t^{\tau-s}\, [t^2 + At + B]^{-\tau-\frac{1}{2}}\, dt \\
&= \frac{(2\pi)^{\tau+\frac{1}{2}}}{\Gamma(\tau+\frac{1}{2})} \int_0^\infty t^{\tau-s}\, dt \int_0^\infty e^{-2\pi v(t^2 + At + B)}\, v^{\tau-\frac{1}{2}}\, dv \\
&= \frac{(2\pi)^{\tau+\frac{1}{2}}}{\Gamma(\tau+\frac{1}{2})} \int_0^\infty t^{-s-\frac{1}{2}}\, dt \int_0^\infty e^{-2\pi v(t + A + Bt^{-1})}\, v^{\tau-\frac{1}{2}}\, dv \\
&= \frac{(2\pi)^{\tau+\frac{1}{2}}}{\Gamma(\tau+\frac{1}{2})} \int_0^\infty e^{-2\pi Av}\, v^{\tau-\frac{1}{2}}\, dv \int_0^\infty t^{-s-\frac{1}{2}}\, e^{-2\pi v(t + Bt^{-1})}\, dt \\
&= \frac{2\,(2\pi)^{\tau+\frac{1}{2}}}{\Gamma(\tau+\frac{1}{2})}\, B^{\frac{1}{4}-\frac{s}{2}} \int_0^\infty e^{-2\pi Av}\, v^{\tau-\frac{1}{2}}\, K_{s-\frac{1}{2}}(4\pi B^{\frac{1}{2}} v)\, dv\ : \tag{10.35}
\end{aligned}
$$

we have used the usual integral expression [16, p. 85] of modified Bessel functions on the last line.

Hence, recalling (8.4),

$$
\begin{aligned}
G_k^{(1)}(z;\, s) &= \frac{2^{\frac{1}{2}}\, (2\pi)^{s+\tau+\frac{3}{2}}}{\Gamma(s)\,\Gamma(\tau+\frac{3}{2})}\, h_{s,-k}(z) \\
&\quad \times k \sum_{\substack{m,n\geq 0 \\ m-n=k}} \bar{a}_m\, a_n \int_0^\infty e^{-2\pi(m+n+2\kappa)v}\, v^{\tau+\frac{1}{2}}\, K_{s-\frac{3}{2}}(2\pi kv)\, dv\,, \tag{10.36}
\end{aligned}
$$

and $G_k^{(2)}(z;\, s)$, has an entirely similar expression, save for the replacement of $K_{s-\frac{3}{2}}$ by $K_{s+\frac{1}{2}}$.

As an immediate generalization of (9.43), one has, with ϕ as defined in (9.41),

$$\int_{-\frac{1}{2}}^{\frac{1}{2}} \phi(u+iv)\, e^{2i\pi ku}\, du = v^\tau \sum_{\substack{m,n\geq 0 \\ m-n=k}} \bar{a}_m\, a_n\, e^{-2\pi(m+n+2\kappa)v}\,. \tag{10.37}$$

Hence, provided convergence can be ensured,

$$G_k^{(1)}(z; s) = \frac{2^{\frac{1}{2}} (2\pi)^{s+\tau+\frac{3}{2}}}{\Gamma(s) \Gamma(\tau + \frac{3}{2})} h_{s,-k}(z)$$

$$\times k \int_{-\frac{1}{2}}^{\frac{1}{2}} \int_0^\infty \phi(u + iv) e^{2i\pi ku} v^{-\frac{1}{2}} K_{s-\frac{3}{2}}(2\pi kv) \, du \, dv \qquad (10.38)$$

or, using definition (8.9) of the function $c_{s,k}$ and recalling that the strip $\mathcal{D}$ has been defined in Proposition 8.1,

$$G_k^{(1)}(z; s) = \frac{2^{\frac{1}{2}} (2\pi)^{s+\tau+\frac{3}{2}}}{\Gamma(s) \Gamma(\tau + \frac{3}{2})} h_{s,-k}(z) \times k \int_{\mathcal{D}} \phi(w) \, c_{s-1, k}(w) \, d\mu(w)$$

$$= \frac{2^{\frac{1}{2}} (2\pi)^{s+\tau+\frac{3}{2}}}{\Gamma(s) \Gamma(\tau + \frac{3}{2})} h_{s,-k}(z) \times k \, A(s - 1, \phi), \qquad (10.39)$$

where we have used the notation (8.56), here recalled for convenience:

$$A(s, \phi) = \int_{\mathcal{D}} c_{s,k}(z) \, \phi(z) \, d\mu(z)$$

$$= \int_{\Gamma \backslash \Pi} \mathfrak{f}_s(z) \, \phi(z) \, d\mu(z). \qquad (10.40)$$

The formula for $G_k^{(2)}(z; s)$ is just the same, only replacing $c_{s-1, k}$ by $c_{s+1, k}$, or $A(s-1, \phi)$ by $A(s+1, \phi)$. As a consequence of Corollary 8.3, the integral on the right-hand side of (10.39) converges if $\frac{3}{2} < \mathrm{Re}\ s < 2$, whereas the one from the corresponding expression of $G_k^{(2)}(z; s)$ converges if $-\frac{1}{2} < \mathrm{Re}\ s < 0$.

From (10.20), one finds that, if one chooses

$$\frac{3}{2} < \sigma_1 < 2, \qquad -\frac{1}{2} < \sigma_2 < 0, \qquad (10.41)$$

one has

$$F_k^{(1)}(z) = \frac{2}{i} \int_{\sigma_1 - i\infty}^{\sigma_1 + i\infty} \psi(1 - 2s) \frac{2^{\frac{1}{2}} (2\pi)^{s+\tau+\frac{3}{2}}}{\Gamma(s) \Gamma(\tau + \frac{3}{2})} h_{s,-k}(z) \times k \, A(s - 1, \phi) \, ds$$

$$\qquad (10.42)$$

and

$$F_k^{(2)}(z) = \frac{2}{i} \int_{\sigma_2 - i\infty}^{\sigma_2 + i\infty} \psi(1 - 2s) \frac{2^{\frac{1}{2}} (2\pi)^{s+\tau+\frac{3}{2}}}{\Gamma(s) \Gamma(\tau + \frac{3}{2})} h_{s,-k}(z) \times k \, A(s + 1, \phi) \, ds :$$

$$\qquad (10.43)$$

recall (5.11) that the function $s \mapsto \psi(1 - 2s)$ is meromorphic for $\operatorname{Re} s > -1$, with a simple pole at $s = 0$; only, this pole does not contribute any extra term to the integral because of the factor $\Gamma(s)$ from the integrand. The convergence of the integral is ensured by the fact that the function $\psi(1 - 2s)$ is rapidly decreasing as $\operatorname{Im} s \to \infty$, since the Gamma factor is taken care of by (8.15) and the estimate concerning $K_{\sigma - \frac{1}{2} + it}(2\pi ky)$ which immediately follows this reference, together with (8.4) and the first line of (10.40), accompanied with (8.9).

We now move both σ_1 and σ_2 from their present values to some value $\frac{1}{2} - \varepsilon \in]0, \frac{1}{2}[$: later, we shall let ε go to zero. Recall from (8.58) that the poles of $A(s, \phi) = \int_F f_s(z) \phi(z) d\mu(z)$ are to be found among those of $f_s(z)$, which have been inventoried in Theorem 8.2. In the course of moving σ_2 to the value $\frac{1}{2} - \varepsilon$, we do not encounter any pole of the integrand of (10.43).

We do cross poles of the integrand of (10.42), however, when moving σ_1 to the same value: from a look at Theorem 8.2 again, $\frac{3}{2}$ is not a pole, but all points $\frac{1}{2} \pm \frac{i\lambda_j}{2}$ are, and so is the point $\frac{1}{2}$. Now, the points of the first kind come by pairs (changing λ_j to $-\lambda_j$) and since

$$\operatorname{Res}\left[f_{s-1}(z); \frac{1}{2} - \frac{i\mu}{2}\right] = \operatorname{Res}\left[f_s(z); -\frac{1}{2} - \frac{i\mu}{2}\right] = -\operatorname{Res}\left[f_s(z); -\frac{1}{2} + \frac{i\mu}{2}\right]$$
$$(10.44)$$

as a consequence of the second equation (8.53), the poles under review do not contribute to the residue theorem, as applied to our case. So far as the point $\frac{1}{2}$ is concerned, let us note that it is a zero of the function $\psi(1 - 2s)$ according to (10.17), which takes care of the fact that it is a simple pole of the function $A(s - 1, \phi)$.

We have reached the point where we can replace both σ_1 and σ_2, as they occur in (10.42) and (10.43), by the common value $\frac{1}{2} - \varepsilon$. We now replace ε by 0, as made possible by the last assertion of Theorem 8.2. The net result is

$$F_k(z) = \frac{2}{i} \int_{\frac{1}{2} - i\infty}^{\frac{1}{2} + i\infty} \psi(1 - 2s) 2^{\frac{1}{2}} \frac{(2\pi)^{s + \tau + \frac{3}{2}}}{\Gamma(s) \Gamma(\tau + \frac{3}{2})} h_{s, -k}(z)$$

$$\times k \left[\int_{\Gamma \backslash \Pi} [f_{s-1}(z) - f_{s+1}(z)] \phi(z) d\mu(z) \right] ds \quad (10.45)$$

or, finally, using (8.18),

$$F_k(z) = \frac{2^{\frac{5}{2}}}{i} \int_{\frac{1}{2} - i\infty}^{\frac{1}{2} + i\infty} \left(\frac{1}{2} - s\right) \psi(1 - 2s) \frac{(2\pi)^{s + \tau + \frac{1}{2}}}{\Gamma(s) \Gamma(\tau + \frac{3}{2})} h_{s, -k}(z)$$

$$\times \left[\int_{\Gamma \backslash \Pi} E_k(z, s) \phi(z) d\mu(z) \right] ds, \quad (10.46)$$

an expression which may be compared to (10.29), rewritten (recall (9.46)) as

$$F_0(z) = \frac{2^{\frac{1}{2}}}{i} \int_{\frac{1}{2}-i\infty}^{\frac{1}{2}+i\infty} \psi(1-2s)\,(2\pi)^{s+\tau+\frac{1}{2}}\,\frac{\Gamma(s-\frac{1}{2})\,\Gamma(\frac{3}{2}-s)}{\Gamma(s)\,\Gamma(\tau+\frac{3}{2})}$$

$$\times \left[y^{1-s} + y^s\,\frac{\zeta^*(2s)}{\zeta^*(2-2s)} \right] \left[\int_{\Gamma\backslash\Pi} E(z,\,s)\,\phi(z)\,d\mu(z) \right]\,ds$$

$$+ \text{Extra}. \quad (10.47)$$

Recall that $\sum_{k\in\mathbb{Z}} F_k(z) = (\mathcal{A}_\tau\,h)(z)$, as defined in (10.3), and set

$$\langle E_s,\,\phi\rangle = \int_{\Gamma\backslash\Pi} E(z,\,s)\,\phi(z)\,d\mu(z),$$

$$\langle E_s^k,\,\phi\rangle = \int_{\Gamma\backslash\Pi} E_k(z,\,s)\,\phi(z)\,d\mu(z). \quad (10.48)$$

Equations (10.46) and (10.47) provide a spectral decomposition of the function $\mathcal{A}_\tau\,h$ reducing to a continuous part, apart from the extra term Extra. According to the Roelcke-Selberg theorem, it can be written as $\text{Extra} + \int_{\frac{1}{2}-i\infty}^{\frac{1}{2}+i\infty} \Psi(s)\,E^*(z,\,s)\,ds$ for some unique density Ψ such that $\Psi(s) = \Psi(1-s)$ (cf. end of proof of Theorem 5.1). Remarking that $y^{1-s} + y^s\,\frac{\zeta^*(2s)}{\zeta^*(2-2s)}$ is the "constant" term of the function $\frac{E^*(z,\,s)}{\zeta^*(2-2s)}$, we obtain the relation

$$\Psi(s) = \frac{2^{\frac{1}{2}}}{i}\,\psi(1-2s)\,(2\pi)^{s+\tau+\frac{1}{2}}\,\frac{\Gamma(s-\frac{1}{2})\,\Gamma(\frac{3}{2}-s)}{\Gamma(s)\,\Gamma(\tau+\frac{3}{2})}\,\frac{\langle E_s,\,\phi\rangle}{\zeta^*(2-2s)} \quad (10.49)$$

provided we show that the function on the right-hand side is invariant under the transformation $s \mapsto 1-s$. According to (10.18), the function $s \mapsto (2\pi)^s\,\psi(1-2s)\frac{\Gamma(s-\frac{1}{2})\,\Gamma(\frac{3}{2}-s)}{\Gamma(s)}$ satisfies the desired invariance property, and so does the other factor $\frac{\langle E_s,\,\phi\rangle}{\zeta^*(2-2s)}$.

Recall the classical Fourier expansion of Eisenstein series [12, p. 68]

$$E^*(z,\,s) = \zeta^*(2-2s)\,y^{1-s} + y^s\,\zeta^*(2s) + 2\sum_{k\neq 0} |k|^{s-\frac{1}{2}}\,\sigma_{1-2s}(|k|)\,h_{s,-k}(z) \quad (10.50)$$

(with $h_{s,k} = h_{s,-k}$ as defined in (8.4); the meaning of $\sigma_{1-2s}(|k|)$ has been recalled in Remark 8.1). As a consequence, for $k > 0$, one has

$$F_k(z) = \frac{2^{\frac{1}{2}}}{i} \int_{\frac{1}{2}-i\infty}^{\frac{1}{2}+i\infty} \psi(1-2s)\,(2\pi)^{s+\tau+\frac{1}{2}}\,\frac{\Gamma(s-\frac{1}{2})\,\Gamma(\frac{3}{2}-s)}{\Gamma(s)\,\Gamma(\tau+\frac{3}{2})}$$

$$\frac{\langle E_s,\,\phi\rangle}{\zeta^*(2-2s)} \times 2\,k^{s-\frac{1}{2}}\,\sigma_{1-2s}(k)\,h_{s,-k}(z)\,ds, \quad (10.51)$$

an expression which has to be identical to (10.46).

Using (8.18) and (8.44), one has

$$E_k(z, 1-s) = \frac{2\pi k}{2s-1}\left[\mathfrak{f}_{-s}(z) - \mathfrak{f}_{2-s}(z)\right]$$

$$= \frac{2\pi k}{2s-1}\left[\mathfrak{f}_{s+1}(z) - \mathfrak{f}_{s-1}(z)\right] = E_k(z, s).\tag{10.52}$$

Looking at (10.46) and (10.51), we observe that the integrands in both decompositions are invariant under the transformation $s \mapsto 1-s$. It follows that these two decompositions into (generalized) eigenfunctions of Δ coincide: hence, for every s ("almost" is unnecessary, since all functions under consideration are continuous on the line $\frac{1}{2} + i\mathbb{R}$, cf. Theorem 8.2), one has

$$\langle E_s^k, \phi \rangle = \frac{1}{2}\Gamma\left(s - \frac{1}{2}\right)\Gamma\left(\frac{1}{2} - s\right)\frac{\langle E_s, \phi \rangle}{\zeta^*(2-2s)}k^{s-\frac{1}{2}}\sigma_{1-2s}(k).\tag{10.53}$$

The value of the function $\alpha_k(s)$ follows:

$$\alpha_k(s) = \frac{1}{2}\frac{\Gamma(s-\frac{1}{2})\Gamma(\frac{1}{2}-s)}{\zeta^*(2-2s)}k^{s-\frac{1}{2}}\sigma_{1-2s}(k).\tag{10.54}$$

Using (9.45) again, we obtain the equation

$$(\mathcal{A}_\tau h)(z) = \text{Extra}$$

$$+ \frac{2\pi^{\frac{1}{2}}}{i}\int_{\frac{1}{2}-i\infty}^{\frac{1}{2}+i\infty}\psi(1-2s)\frac{\Gamma(s-\frac{1}{2})\Gamma(\frac{3}{2}-s)\Gamma(s+\tau)}{2^{s+\tau}\,\Gamma(s)\,\Gamma(\tau+\frac{3}{2})}\frac{L(\overline{f}\otimes f, s)}{\zeta^*(2-2s)}E^*(z, s)\,ds,\tag{10.55}$$

with Extra as given in (10.27).

It has been necessary for the proof, starting from the left-hand side of (10.7), to make the assumption (10.5) so that ψ should vanish at 0 (10.17) but, as it turns out, this assumption is far from necessary for the right-hand side to be meaningful since on one hand $\zeta^*(2-2s)$ has a pole at $s = \frac{1}{2}$, on the other hand, as it follows from (9.46), the function $L(\overline{f}\otimes f, s)$ vanishes at $s = \frac{1}{2}$. The necessity of this assumption arose from the singularity of each of the two terms $A(s\pm 1, \phi)$ of (10.42) and (10.43), when taken individually, while their sum vanishes at $s = \frac{1}{2}$ in view of (8.44). □

For future reference, note that (10.55) can be rewritten as

$$(\mathcal{A}_\tau h)(z) = -2^{-\tau-1}\pi^{\frac{1}{2}}\frac{\Gamma(\tau+1)}{\Gamma(\tau+\frac{3}{2})}h(0)\,\text{Res}_{s=1}\left(L(\overline{f}\otimes f, s)\right)\tag{10.56}$$

$$+ \pi^{\frac{1}{2}}\int_{-\infty}^{\infty}\psi(i\lambda)\frac{\Gamma(-\frac{i\lambda}{2})\Gamma(1+\frac{i\lambda}{2})\Gamma(\frac{1-i\lambda}{2}+\tau)}{2^{\frac{1-i\lambda}{2}+\tau}\Gamma(\frac{1-i\lambda}{2})\Gamma(\tau+\frac{3}{2})}\frac{L(\overline{f}\otimes f, \frac{1-i\lambda}{2})}{\zeta^*(-i\lambda)}E^*\left(z, \frac{1-i\lambda}{2}\right)d\lambda.$$

Corollary 10.3. *Under the assumptions of Theorem 10.1, and assuming more-over that $h(0) = 0$, let h^{iso} be the isometric horocyclic symbol of the operator $\mathrm{Op}^{\tau}_{\mathrm{soft}}(h)$. One has*

$$\| \mathcal{A}_{\tau} h \|^2_{L^2(\Gamma \backslash \Pi)} = \frac{1}{\pi} \left\| \frac{\zeta(1 - 2i\pi\mathcal{E})}{\zeta(-2i\pi\mathcal{E})} \, L\left(\overline{f} \otimes f, \frac{1}{2} - i\pi\mathcal{E} \right) h^{\mathrm{iso}} \right\|^2_{L^2(\mathbb{R}^2)}. \qquad (10.57)$$

Proof. It is entirely similar to that of Corollary 5.2. In view of the symmetry condition following (10.49), the identity (10.56) expresses the proper Roelcke-Selberg expansion of the function $\mathcal{A}_{\tau} h$. Hence, just as in (5.31), one has

$$\int_{\Gamma \backslash \Pi} \overline{E\left(z, \frac{1 - i\lambda}{2} \right)} (\mathcal{A}_{\tau} h)(z) \, d\mu(z)$$

$$= 2^{\frac{5+i\lambda}{2} - \tau} \frac{\Gamma(1 + \frac{i\lambda}{2}) \Gamma(\frac{1-i\lambda}{2} + \tau)}{\Gamma(\tau + \frac{3}{2})} \frac{\zeta(1 - i\lambda)}{\zeta(-i\lambda)} L\left(\overline{f} \otimes f, \frac{1 - i\lambda}{2} \right) \psi(i\lambda) : \qquad (10.58)$$

it follows that

$$\| \mathcal{A}_{\tau} h \|^2_{L^2(\Gamma \backslash \Pi)} \qquad\qquad\qquad\qquad\qquad\qquad\qquad\qquad (10.59)$$

$$= \frac{2^{-2\tau}}{\pi} \left\| \frac{\Gamma(1 + i\pi\mathcal{E}) \Gamma(\frac{1}{2} + \tau - i\pi\mathcal{E})}{\Gamma(\tau + \frac{3}{2})} \frac{\zeta(1 - 2i\pi\mathcal{E})}{\zeta(-2i\pi\mathcal{E})} L\left(\overline{f} \otimes f, \frac{1}{2} - i\pi\mathcal{E} \right) h^{\mathrm{iso}} \right\|^2_{L^2(\mathbb{R}^2)}.$$

What remains to be done is only substituting for the "hard" symbol h the right-hand side of the equation

$$h = \frac{2^{-\tau}}{\Gamma(\tau + \frac{3}{2})} \Gamma(1 + i\pi\mathcal{E}) \Gamma\left(\frac{1}{2} + \tau - i\pi\mathcal{E} \right) h^{\mathrm{iso}}, \qquad (10.60)$$

as provided by (7.37), not forgetting that the link from the hard symbol to the isometric (horocyclic) symbol of some operator is provided by the adjoint of the operator linking the soft symbol to the isometric symbol.   □

11 Beyond the radial case: automorphic distributions

One of the drawbacks of Theorem 10.2 is that it depends in an essential way on the fact that h is a radial symbol. Indeed, if this condition is dropped, equation (10.3) will produce a function on $SL(2, \mathbb{Z}) \backslash SL(2, \mathbb{R})$, not on the double quotient $SL(2, \mathbb{Z}) \backslash SL(2, \mathbb{R}) / SO(2)$. One may still avoid the use of the first homogeneous space, substituting for it the quotient of $\mathbb{R}^2 \backslash \{0\}$ by the linear action of $SL(2, \mathbb{Z})$: on this rather singular space (most orbits are everywhere dense), distributions, called *automorphic distributions*, rather than functions, make perfectly good sense. With this concept in mind, our new formulation will be an extension, in the τ-calculus, of the result of Section 4 (relative to the Weyl calculus and associated

Wigner functions) rather than that in Section 5. Our detour through automorphic function theory, however, was necessary, in view of the rather severe analytical difficulties of the proof of Theorem 10.1.

In [30, Section 13] and [31], we considered automorphic distributions in the plane, by which we meant Γ-invariant tempered distributions on $\mathbb{R}^2$, *i.e.*, distributions $\mathfrak{S} \in \mathcal{S}'(\mathbb{R}^2)$ with the property that $\langle \mathfrak{S}, h \circ g \rangle = \langle \mathfrak{S}, h \rangle$ for every $h \in \mathcal{S}(\mathbb{R}^2)$ and every $g \in \Gamma = SL(2, \mathbb{Z})$. It was shown in *loc. cit.* that, with the help of the Weyl calculus, automorphic distributions can be transformed into pairs (f_0, f_1) of automorphic functions in Π, linking the notion to the Lax-Phillips scattering theory [15] for the automorphic wave equation. The advantages of automorphic distribution theory, as opposed to non-holomorphic modular form theory, are the following: first, it is often easier to decompose functions or distributions on $\mathbb{R}^2$ into their homogeneous components than to decompose functions in Π into (generalized) eigenfunctions of Δ. Next, the notion is *slightly* subtler: for instance, the two modular forms $E(z, \frac{1 \pm \nu}{2})$ are proportional, while the distributions $\mathfrak{E}_{\pm\nu}$ (*cf. infra*) are related by the symplectic Fourier transformation instead. One last advantage has to do with covariance, and has been explained in the beginning of the present section: it only requires that we consider $\mathbb{R}^2$ as a linear space without any given, or assumed, Euclidean structure; in this case, of course, the correspondence $\mathfrak{S} \mapsto (f_0, f_1)$ ceases to be canonical. A calculus of operators with automorphic distributions as Weyl-type symbols constituted the bulk of [31].

Here, our interest in automorphic distributions has another, somewhat dual, origin: automorphic distributions appear as Wigner functions, in the calculus Op^τ, of the kind $W^\tau(\mathfrak{s}_\tau, \mathfrak{s}_\tau)$ as tested against radial symbols in (10.3), and we wish to make these more explicit. An essential difference (a simplification) in the present context is that we shall not have to consider what we called cusp-distributions in the former one (those automorphic distributions which give rise to pairs of non-holomorphic cusp-forms under the correspondence alluded to above), only Eisenstein distributions. Let us recall how these are defined [30, Prop. 13.1] or [31, p. 18]. For $\mathrm{Re}\ \nu < -1$, $h \in \mathcal{S}(\mathbb{R}^2)$, set

$$\langle \mathfrak{E}_\nu, h \rangle = \frac{1}{2} \sum_{|m|+|n|\neq 0} \int_{-\infty}^{\infty} |t|^{-\nu} h(tn, tm)\, dt, \qquad (11.1)$$

a convergent expression. In the references above, we denoted this as $< \mathfrak{E}_\nu^\sharp, h >$, but there is no reason here to keep the superscript $\sharp$. This defines an even tempered distribution $\mathfrak{E}_\nu$, homogeneous of degree $-1-\nu$, and the function $\nu \mapsto \mathfrak{E}_\nu$ extends as a holomorphic function of ν for $\nu \neq \pm 1$, with simple poles at $\nu = \pm 1$; the residues there are given as $\mathrm{Res}_{\nu=-1}\, \mathfrak{E}_\nu = -1$ and $\mathrm{Res}_{\nu=1}\, \mathfrak{E}_\nu = \delta$, the unit mass at the origin of $\mathbb{R}^2$. Also, $\mathcal{F}\mathfrak{E}_\nu = \mathfrak{E}_{-\nu}$. We shall also use, here, a different normalization, setting

$$\mathfrak{F}_\nu = 2^{\frac{-1-\nu}{2}}\, \mathfrak{E}_\nu : \qquad (11.2)$$

since $\mathfrak{E}_\nu$ is homogeneous of degree $-1 - \nu$, this can be written

$$\mathfrak{F}_\nu = 2^{-\frac{1}{2}+i\pi\mathcal{E}}\, \mathfrak{E}_\nu\,. \tag{11.3}$$

Then,

$$\mathcal{G}\mathfrak{F}_\nu = \mathfrak{F}_{-\nu}\,, \tag{11.4}$$

with $\mathcal{G}$ as defined in (2.8). The fundamental role of Eisenstein distributions lies in the equation (*cf.* Proposition 4.1 for the definition of the Dirac comb $\mathfrak{D}_0$)

$$\mathfrak{D}_0 = \delta + 1 + \frac{1}{2\pi}\int_{-\infty}^{\infty}\mathfrak{E}_{i\lambda}\,d\lambda \tag{11.5}$$

which, apart from two extra terms away from the spectral line (δ is a homogeneous distribution of degree -2, while the constant 1 is of course homogeneous of degree 0), expresses $\mathfrak{D}_0$ as an integral superposition of Eisenstein distributions. In [32], a decomposition into homogeneous components of related automorphic distributions led to some generalizations of Poisson's formula, in the spirit of Voronoi's formula [34].

Note that, by definition, an automorphic distribution is even, since the matrix $-I$ lies in Γ. It does not have to be $\mathcal{G}$-invariant or antiinvariant but, in the present context (in contradiction to our use in the former reference), we shall mostly use such species of automorphic distributions. The reason for this is that the Weyl calculus on the line, when restricted to even symbols, in other words to operators which commute with the map $u \mapsto \check{u}$, $\check{u}(t) = u(-t)$, is not equivalent to just one Op^τ-calculus as considered in the present work, rather to the direct sum of two such calculi, those corresponding to $\tau = \pm\frac{1}{2}$: on the other hand, we here consider only one such calculus. In order to make the link between Eisenstein distributions and Eisenstein series explicit, we need a lemma, consisting of the first part of the following proposition.

Proposition 11.1. *The isometric horocyclic symbol h^{iso} of the projection operator $u \mapsto (\psi_i^{\tau+1}\,|\,u)_{\tau+1}\,\psi_i^{\tau+1}$ has its homogeneous components given by the equation*

$$(h^{\mathrm{iso}})_\lambda(q,\,p) = 2^{-\frac{1}{2}}\,(2\pi)^{-1-\frac{i\lambda}{2}}\,\frac{\Gamma(\frac{1+i\lambda}{2})\,\Gamma(\tau + \frac{1+i\lambda}{2})}{\Gamma(\frac{i\lambda}{2})\,\Gamma(\tau+1)}\,(q^2 + p^2)^{\frac{-1-i\lambda}{2}}\,. \tag{11.6}$$

In the case when $\tau = -\frac{1}{2}$ (resp. $\tau = \frac{1}{2}$), the isometric horocyclic symbol of an operator A in $H_{\tau+1}$ coincides with the Weyl symbol of the transfer under $\mathrm{Sq}_{\mathrm{even}}$ (resp. $\mathrm{Sq}_{\mathrm{odd}}$) of A, as an even-even (resp. odd-odd) operator.

Proof. We start with recalling from [26, p. 106] that the passive symbol of the operator under consideration is the function

$$f^{\mathrm{pass}}(z) = 2\,(\cosh d(i,\,z))^{-\tau-1}\,. \tag{11.7}$$

The Mehler inversion formula [16, p. 398] or [24, p. 144] makes it possible to write

$$\delta^{-\tau-1} = \int_0^\infty k(\lambda)\, \mathfrak{P}_{-\frac{1}{2}+\frac{i\lambda}{2}}(\delta)\, d\lambda \qquad (11.8)$$

with

$$k(\lambda) = \frac{\lambda}{4}\left(\tanh\frac{\pi\lambda}{2}\right) \int_1^\infty \mathfrak{P}_{-\frac{1}{2}+\frac{i\lambda}{2}}(\delta)\, \delta^{-\tau-1}\, d\delta. \qquad (11.9)$$

This integral can be made explicit [24, p. 146] or [10, p.183], which yields

$$f^{\mathrm{pass}}(z) = \int_0^\infty \phi(\lambda)\, \mathfrak{P}_{-\frac{1}{2}+\frac{i\lambda}{2}}(\cosh d(i, z))\, d\lambda \qquad (11.10)$$

with

$$\phi(\lambda) = \frac{\Gamma(\frac{1+i\lambda}{2})\Gamma(\frac{1-i\lambda}{2})}{\Gamma(\frac{i\lambda}{2})\Gamma(-\frac{i\lambda}{2})}\, \frac{\pi^{-\frac{1}{2}}}{\Gamma(\tau+1)}\, 2^{\tau-1}\, \Gamma\left(\frac{\tau}{2} + \frac{1+i\lambda}{4}\right) \Gamma\left(\frac{\tau}{2} + \frac{1-i\lambda}{4}\right). \qquad (11.11)$$

Equation (7.3) gives the isometric horocyclic symbol h^{iso} in terms of the active symbol f^{act}, (6.31) gives f^{pass} in terms of f^{act} and, under the transformation TV introduced in the beginning of Section 7, the operators $\Delta - \frac{1}{4}$ on Π and $\pi^2 \mathcal{E}^2$ on $\mathbb{R}^2$ correspond. One can then write

$$h^{\mathrm{iso}} = (2\pi)^{-\frac{1}{2}}\, \pi^{i\pi\mathcal{E}}\, \frac{\Gamma(\frac{\tau}{2} + \frac{3}{4} - \frac{i\pi\mathcal{E}}{2})}{\Gamma(\frac{\tau}{2} + \frac{3}{4} + \frac{i\pi\mathcal{E}}{2})}\, TV\, f^{\mathrm{pass}}. \qquad (11.12)$$

Applying (11.10), we must now compute the transform, under TV, of the function $z \mapsto \mathfrak{P}_{-\frac{1}{2}+\frac{i\lambda}{2}}(\cosh d(i, z))$: since the latter function is both radial and a generalized eigenfunction of Δ for the eigenvalue $\frac{1+\lambda^2}{4}$, its TV-transform is a linear combination of the functions $(q, p) \mapsto (q^2 + p^2)^{-\frac{1}{2}\pm\frac{i\lambda}{2}}$. One has [30, p. 30]

$$\mathfrak{P}_{-\frac{1}{2}+\frac{i\lambda}{2}}(\cosh d(i, z)) = \frac{1}{\pi} \int_{-\infty}^\infty \left(\frac{|z-s|^2}{\mathrm{Im}\, z}\right)^{-\frac{1}{2}+\frac{i\lambda}{2}} (s^2 + 1)^{-\frac{1}{2}-\frac{i\lambda}{2}}\, ds. \qquad (11.13)$$

With the help of the adjoint V^*T^* of TV, this can also be written (loc. cit.) as

$$\mathfrak{P}_{-\frac{1}{2}+\frac{i\lambda}{2}}(\cosh d(i, z)) = \left(\frac{2}{\pi}\right)^{\frac{1}{2}} V^*T^*\left[(q, p) \mapsto \frac{\Gamma(-\frac{i\lambda}{2})}{\Gamma(\frac{1-i\lambda}{2})}\, (q^2 + p^2)^{-\frac{1}{2}-\frac{i\lambda}{2}}\right](z). \qquad (11.14)$$

Now [30, Theorem 4.1], V^*T^* is a left-inverse, but not a right-inverse, of TV: the product TVV^*T^*, applied to some even function on $\mathbb{R}^2$, yields the part of this function invariant under the involution denoted as $T\kappa T^{-1}$ in the reference just given: from [30, p. 26], one has

$$T\kappa T^{-1} = (2\pi)^{-2i\pi\mathcal{E}}\, \frac{\Gamma(i\pi\mathcal{E})}{\Gamma(-i\pi\mathcal{E})}\, \mathcal{G}. \qquad (11.15)$$

Using the Hecke (or Bochner) equation (9.12) for the Fourier transform of a radial function, one obtains

$$\mathcal{F}\left((q,\,p)\mapsto (q^2+p^2)^{-\frac{1}{2}-\frac{i\lambda}{2}}\right)=(2\pi)^{i\lambda}\,\frac{\Gamma(\frac{1-i\lambda}{2})}{\Gamma(\frac{1+i\lambda}{2})}\,(q^2+p^2)^{-\frac{1}{2}+\frac{i\lambda}{2}}\;:\qquad(11.16)$$

since, on functions homogeneous of degree $-1-i\lambda$, the operator $2i\pi\mathcal{E}$ coincides with the multiplication by $-i\lambda$, one obtains

$$T\kappa T^{-1}\left[\frac{\Gamma(-\frac{i\lambda}{2})}{\Gamma(\frac{1-i\lambda}{2})}\,(q^2+p^2)^{-\frac{1}{2}-\frac{i\lambda}{2}}\right]=\frac{\Gamma(\frac{i\lambda}{2})}{\Gamma(\frac{1+i\lambda}{2})}\,(q^2+p^2)^{-\frac{1}{2}+\frac{i\lambda}{2}}\,.\qquad(11.17)$$

It then follows from (11.14) that, whether $\lambda>0$ or $\lambda<0$, one has

$$TV\left[z\mapsto\mathfrak{P}_{-\frac{1}{2}+\frac{i\lambda}{2}}(\cosh d(i,\,z))\right](q,\,p)$$

$$=(2\pi)^{-\frac{1}{2}}\left[\frac{\Gamma(-\frac{i\lambda}{2})}{\Gamma(\frac{1-i\lambda}{2})}\,(q^2+p^2)^{-\frac{1}{2}-\frac{i\lambda}{2}}+\frac{\Gamma(\frac{i\lambda}{2})}{\Gamma(\frac{1+i\lambda}{2})}\,(q^2+p^2)^{-\frac{1}{2}+\frac{i\lambda}{2}}\right]\,.\qquad(11.18)$$

Equation (11.6) follows.

For the second part, let us recall that, with the notation of Section 2, the function u_i (*resp.* u_i^1) on the real line is the transfer under $\mathrm{Sq}_{\mathrm{even}}$ (*resp.* $\mathrm{Sq}_{\mathrm{odd}}$) of the function $\psi_i^{\frac{1}{2}}$ (*resp.* $\psi_i^{\frac{3}{2}}$) on the half-line. In the Weyl calculus, the symbols $W(u_i,\,u_i)$ and $W(u_i^1,\,u_i^1)$ of the orthogonal projection operators on u_i or u_i^1 are given, as indicated in (2.10), (2.11), by the equations

$$W(u_i,\,u_i)(q,\,p)=\Phi_1(q,\,p)\,,$$

$$W(u_i^1,\,u_i^1)(q,\,p)=\left(-2\frac{d}{dt}-1\right)\bigg|_{t=1}\Phi_t(q,\,p)\,,\qquad(11.19)$$

with

$$\Phi_t(q,\,p)=2\,e^{-2\pi t(q^2+p^2)}\,.\qquad(11.20)$$

It is immediate, using (7.9), that

$$(\Phi_t)_\lambda(q,\,p)=(2\pi)^{\frac{-3-i\lambda}{2}}\,\Gamma\left(\frac{1+i\lambda}{2}\right)(q^2+p^2)^{\frac{-1-i\lambda}{2}}\,t^{\frac{-1-i\lambda}{2}}\;:\qquad(11.21)$$

the coincidence between $W(u_i,\,u_i)$ (*resp.* $W(u_i^1,\,u_i^1)$) and the isometric horocyclic symbol of the corresponding projection operator in $H_{\frac{1}{2}}$ (*resp.* $H_{\frac{3}{2}}$) follows. Using the covariance of the Weyl calculus (*resp.* the τ-calculus) under the metaplectic representation (*resp.* the representation $\mathcal{D}_{\tau+1}$), one extends the coincidence just obtained to the case when u_i (*resp.* u_i^1) is replaced by u_z (*resp.* u_z^1) and $\psi_i^{\tau+1}$ is replaced by $\psi_z^{\tau+1}$. A sesquiholomorphic argument, based on the fact that $\psi_z^{\tau+1}$ is the product of $(\text{Im}\,(-z^{-1}))^{\frac{\tau+1}{2}}$ by an antiholomorphic function of z and a related one concerning the metaplectic representation, shows that the coincidence under study extends to the case when, instead of a projection operator, one considers the rank-one operator based on the introduction of a pair $(\psi_w^{\tau+1},\,\psi_z^{\tau+1})$ in place of a diagonal one. Lemma 11.1 follows, with the help of an obvious density argument. □

Theorem 11.2. *Let* $\Theta_\tau : \mathfrak{S} \mapsto f$ *be the linear map from* $\mathcal{G}$-*antiinvariant automorphic distributions to automorphic functions defined by the equation*

$$f(z) = (\psi_z^{\tau+1} \,|\, \mathrm{Op}_{\mathrm{soft}}^\tau(\mathfrak{S})\,\psi_z^{\tau+1}), \tag{11.22}$$

with $\psi_z^{\tau+1}$ *as defined in (6.6). One has, for* $\nu \neq \pm(2\tau+1+2k)$, $k = 0, 1, \dots$,

$$\left(\Theta_\tau\,\frac{\mathfrak{F}_\nu - \mathfrak{F}_{-\nu}}{2}\right)(z) = -2^{-\tau-1}\,\pi^{\frac{1}{2}}\,\nu\,\frac{\Gamma(\tau+\frac{1-\nu}{2})\,\Gamma(\tau+\frac{1+\nu}{2})}{\Gamma(\tau+\frac{3}{2})\,\Gamma(\tau+1)}\,E^*\left(z,\,\frac{1-\nu}{2}\right). \tag{11.23}$$

Proof. From Definition 7.3 and the preceding proposition, it follows that the components of the soft horocyclic symbol h^{soft} of the projection operator $P_{\psi_i^{\tau+1},\,\psi_i^{\tau+1}}$ are given by the equation

$$h_\lambda^{\mathrm{soft}}(q,\,p) = \frac{2^{-\tau}}{\Gamma(\tau+\frac{3}{2})}\,\Gamma\left(1+\frac{i\lambda}{2}\right)\,\Gamma\left(\tau+\frac{1-i\lambda}{2}\right)\,h_\lambda^\flat$$

$$= c_{i\lambda}\,(q^2+p^2)^{\frac{-1-i\lambda}{2}} \tag{11.24}$$

with

$$c_{i\lambda} = 2^{-\tau-\frac{1}{2}}\,\frac{i\lambda}{2}\,(2\pi)^{-1-\frac{i\lambda}{2}}\,\frac{\Gamma(\frac{1+i\lambda}{2})\,\Gamma(\tau+\frac{1+i\lambda}{2})\,\Gamma(\tau+\frac{1-i\lambda}{2})}{\Gamma(\tau+\frac{3}{2})\,\Gamma(\tau+1)}. \tag{11.25}$$

According to (7.58), one has

$$\left(\Theta_\tau\,\frac{\mathfrak{F}_\nu - \mathfrak{F}_{-\nu}}{2}\right)(i) = 2^{\frac{-1-\nu}{2}}\,\langle\,\mathfrak{E}_\nu,\,h^{\mathrm{soft}}\,\rangle : \tag{11.26}$$

note that, when applying (7.58), we have traded the $\mathcal{G}$-antiinvariant distribution $\frac{\mathfrak{F}_\nu - \mathfrak{F}_{-\nu}}{2}$ for the simpler distribution $\mathfrak{F}_\nu$; this does not change anything since, anyway, the function h^{soft} is itself $\mathcal{G}$-antiinvariant. Using (11.1) and (11.2), one finds

$$(\Theta_\tau\,\frac{\mathfrak{F}_\nu - \mathfrak{F}_{-\nu}}{2})(i) = 2^{\frac{-1-\nu}{2}}\,\langle\,\mathfrak{E}_\nu,\,h^{\mathrm{soft}}\,\rangle$$

$$= 2^{\frac{-1-\nu}{2}}\,\langle\,\mathfrak{D}_0 - \delta,\,(x,\,\xi) \mapsto \int_0^\infty t^{-\nu}\,h^{\mathrm{soft}}(tx,\,t\xi)\,dt\,\rangle. \tag{11.27}$$

Now, from (7.9), one has

$$\int_0^\infty t^{i\lambda}\, h^{\text{soft}}(tx,\, t\xi)\, dt = 2\pi\, c_{i\lambda}\, (x^2 + \xi^2)^{\frac{-1-i\lambda}{2}}. \tag{11.28}$$

Defining $c_{-\nu}$, for $\operatorname{Re}\nu < -1$ and ν distinct from the values indicated in the statement of Theorem 10.2, by analytic continuation from $c_{i\lambda}$ as defined in (11.25), one obtains

$$\int_0^\infty t^{-\nu}\, h^{\text{soft}}(tx,\, t\xi)\, dt = 2\pi\, c_{-\nu}\, (x^2 + \xi^2)^{\frac{\nu-1}{2}}. \tag{11.29}$$

Finally,

$$\left(\Theta_\tau\, \frac{\mathfrak{F}_\nu - \mathfrak{F}_{-\nu}}{2}\right)(i) = 2^{\frac{-1-\nu}{2}} \cdot 2\pi\, c_{-\nu}\, \left\langle \mathfrak{D}_0 - \delta,\, (x,\,\xi) \mapsto (x^2 + \xi^2)^{\frac{\nu-1}{2}} \right\rangle$$

$$= 2^{\frac{1-\nu}{2}}\, \pi\, c_{-\nu} \sum_{|m|+|n|\neq 0} (m^2 + n^2)^{\frac{\nu-1}{2}}$$

$$= 2^{\frac{3-\nu}{2}}\, \pi\, c_{-\nu}\, \zeta(1-\nu)\, E\left(i,\, \frac{1-\nu}{2}\right)$$

$$= 2^{\frac{3-\nu}{2}}\, \pi\, c_{-\nu}\, \frac{\pi^{\frac{1-\nu}{2}}}{\Gamma(\frac{1-\nu}{2})}\, E^*\left(i,\, \frac{1-\nu}{2}\right): \tag{11.30}$$

using the covariance of the $\operatorname{Op}^\tau_{\text{soft}}$-calculus, one has

$$\left(\Theta_\tau\, \frac{\mathfrak{F}_\nu - \mathfrak{F}_{-\nu}}{2}\right)(z) = 2^{\frac{3-\nu}{2}}\, \pi\, c_{-\nu}\, \frac{\pi^{\frac{1-\nu}{2}}}{\Gamma(\frac{1-\nu}{2})}\, E^*\left(z,\, \frac{1-\nu}{2}\right) \tag{11.31}$$

for every $z \in \Pi$. This leads to the equation which is the point of Theorem 11.2.

The following corollary, which expresses the same result in terms of the isometric horocyclic calculus Op^τ in place of the soft horocyclic calculus $\operatorname{Op}^\tau_{\text{soft}}$, is meant mostly as a means of verification. Note that, so as to obtain a symbol satisfying the τ-dependent symmetry property (7.5), we must consider this time the automorphic distribution

$$\mathfrak{T}_\nu = \frac{1}{2}\left[\mathfrak{F}_\nu + \frac{\Gamma(\frac{\nu}{2})\Gamma(\tau + \frac{1-\nu}{2})}{\Gamma(-\frac{\nu}{2})\Gamma(\tau + \frac{1+\nu}{2})}\, \mathfrak{F}_{-\nu}\right]. \tag{11.32}$$

Corollary 11.3. *Under the conditions of Theorem 11.2, and with $\mathfrak{T}_\nu$ as in (11.32), one has*

$$(\psi_z^{\tau+1}\,|\,\operatorname{Op}^\tau(\mathfrak{T}_\nu)\,\psi_z^{\tau+1}) = \pi^{\frac{1}{2}}\, \frac{\Gamma(\tau + \frac{1-\nu}{2})}{\Gamma(-\frac{\nu}{2})\Gamma(\tau + 1)}\, E^*\left(z,\, \frac{1-\nu}{2}\right). \tag{11.33}$$

Proof. If one uses (11.6) instead of (11.24), which amounts to substituting

$$c'_{i\lambda} = 2^{-\frac{1}{2}}\, (2\pi)^{-1-\frac{i\lambda}{2}}\, \frac{\Gamma(\frac{1+i\lambda}{2})\Gamma(\tau + \frac{1+i\lambda}{2})}{\Gamma(\frac{i\lambda}{2})\Gamma(\tau + 1)} \tag{11.34}$$

for $c_{i\lambda}$, the proof is identical to that of Theorem 11.2. $\qquad\square$

Set $A = \mathrm{Op}^\tau(\mathfrak{T}_\nu)$. In the case when $\tau = -\frac{1}{2}$, the formula just obtained reduces to $(\psi_z^{\frac{1}{2}} \mid A \psi_z^{\frac{1}{2}}) = E^*(z, \frac{1-\nu}{2})$: transferring it under the operator $\mathrm{Sq}_{\text{even}}$, it is fully equivalent to the equation [31, p. 20]

$$(u_z \mid \mathrm{Op}(\mathfrak{F}_\nu) u_z) = E^*\left(z, \frac{1-\nu}{2}\right) \tag{11.35}$$

from the Weyl calculus (*cf.* (2.12) for a definition of u_z). The case when $\tau = \frac{1}{2}$ of the present corollary is equivalent to the equation (*loc. cit.*)

$$(u_z^1 \mid \mathrm{Op}(\mathfrak{F}_\nu) u_z^1) = -\nu\, E^*\left(z, \frac{1-\nu}{2}\right) : \tag{11.36}$$

in the two equations, the condition $\nu \neq \pm 1$, which is the one making $\mathfrak{E}_{\pm 1}$ meaningful as a tempered distribution, suffices. □

Though it has been quite useful technically, there is no need to phrase Theorem 10.1, as we have done, in a way dependent on the fact that we are dealing with a radial symbol h. Indeed, we shall give a formulation of this theorem in which it will be possible to substitute for the standard quadratic form $(q, p) \mapsto q^2 + p^2$ a transform of it under the linear change of coordinates associated with a matrix $g \in G$: the present formulation would not make this possible. Then, we must start with showing how an even $\mathcal{G}$-antiinvariant symbol h can be rebuilt as an integral superposition of transforms, under elements of G, of radial $\mathcal{G}$-antiinvariant symbols.

To do this, we shall take advantage of the set of formulas, in Sections 6 and 7, linking the symbols of various species of operators in the space $H_{\frac{3}{2}}$, recalling that this is also the image under $\mathrm{Sq}_{\text{odd}}^{-1}$ of the space $L_{\text{odd}}^2(\mathbb{R}^2)$. Definition 7.1 gives the relation

$$h^{\text{iso}} = \mathcal{R}_{\frac{3}{2}} T V f^{\text{act}}, \tag{11.37}$$

with

$$\mathcal{R}_{\frac{3}{2}} = (2\pi)^{\frac{1}{2}} \pi^{i\pi\mathcal{E}} \frac{\Gamma(\frac{1-i\pi\mathcal{E}}{2})}{\Gamma(\frac{2+i\pi\mathcal{E}}{2})} : \tag{11.38}$$

since, as already mentioned in the proof that precedes, V^*T^* is a left-inverse of TV, and since

$$\mathcal{R}_{\frac{3}{2}}^* \mathcal{R}_{\frac{3}{2}} = 2\pi \frac{\Gamma(\frac{1-i\pi\mathcal{E}}{2})\Gamma(\frac{1+i\pi\mathcal{E}}{2})}{\Gamma(\frac{2+i\pi\mathcal{E}}{2})\Gamma(\frac{2-i\pi\mathcal{E}}{2})} \tag{11.39}$$

transfers, under TV to the operator

$$2\pi \frac{\Gamma\left(\frac{1}{2} - \frac{i}{2}(\Delta - \frac{1}{4})^{\frac{1}{2}}\right) \Gamma\left(\frac{1}{2} + \frac{i}{2}(\Delta - \frac{1}{4})^{\frac{1}{2}}\right)}{\Gamma\left(1 + \frac{i}{2}(\Delta - \frac{1}{4})^{\frac{1}{2}}\right) \Gamma\left(1 - \frac{i}{2}(\Delta - \frac{1}{4})^{\frac{1}{2}}\right)} \tag{11.40}$$

on Π, one has

$$\| h^{\text{iso}} \|_{L^2(\mathbb{R}^2)} = \| \Lambda f^{\text{act}} \|_{L^2(\Pi)} \tag{11.41}$$

with

$$\Lambda = (2\pi)^{\frac{1}{2}} \frac{\Gamma\left(\frac{1}{2} + \frac{i}{2}(\Delta - \frac{1}{4})^{\frac{1}{2}}\right)}{\Gamma\left(1 + \frac{i}{2}(\Delta - \frac{1}{4})^{\frac{1}{2}}\right)} \tag{11.42}$$

and, in the reverse direction,

$$\| f^{\text{act}} \|_{L^2(\Pi)} = (2\pi)^{-\frac{1}{2}} \left\| \frac{\Gamma(\frac{2+i\pi\mathcal{E}}{2})}{\Gamma(\frac{1-i\pi\mathcal{E}}{2})} h^{\text{iso}} \right\|_{L^2(\mathbb{R}^2)} \tag{11.43}$$

(variants are possible: one can pick the other factors above and below). From the first equation, h^{iso} is square-integrable (and, by way of consequence, the associated operator is Hilbert-Schmidt) if f^{act} lies in $L^2(\Pi)$ and is the image under $\Delta - \frac{1}{4}$ of another function in $L^2(\Pi)$. From the second equation, f^{act} lies in $L^2(\Pi)$ if both h^{iso} and the image of that function under $2i\pi\mathcal{E}$ both lie in $L^2(\mathbb{R}^2)$.

We now make the operator $\Theta \colon f^{\text{act}} \mapsto h^{\text{iso}}$ and its inverse explicit (the formulas, without details, were given in [28, p. 271]).

Proposition 11.4. *If $z = g.i$ with $g \in G$, set, starting from the standard norm $\| \ |$ on $\mathbb{R}^2$,*

$$\| (q, p) \|_z^2 = |g^{-1}(q, p)|^2 = \frac{q^2 - 2x\, qp + |z|^2 p^2}{y}. \tag{11.44}$$

For every $f \in C_0^\infty(\Pi)$, the function $h = \Theta f$ is given by the equation

$$h(q, p) = \frac{2^{\frac{3}{2}}}{i} \int_\Pi f(z) \cos\left(2\pi \| (q, p) \|_z^2 + \frac{\pi}{4}\right) d\mu(z). \tag{11.45}$$

The map Θ intertwines the two quasi-regular actions of G on functions defined in Π or in $\mathbb{R}^2$. Also, a fact already used after (2.18), under the map Θ, the operator $\Delta - \frac{1}{4}$ transfers to $\pi^2 \mathcal{E}^2$ [31, p. 17]: this does not depend on the exact form of the integral kernel in (11.45), only on the fact that it is a function of $\| (q, p) \|_z^2$. The image of Θ consists of functions changing to their negatives under the involution $\mathcal{G}$. The inverse map is given by the equation

$$f(z) = 2^{\frac{3}{2}} i \int_{\mathbb{R}^2} h_1(q, p) \cos\left(2\pi \| (q, p) \|_z^2 + \frac{\pi}{4}\right) dq\, dp \tag{11.46}$$

with

$$h_1 = \frac{1}{2\pi} \frac{\Gamma\left(1 + \frac{i\pi\mathcal{E}}{2}\right) \Gamma\left(1 - \frac{i\pi\mathcal{E}}{2}\right)}{\Gamma\left(\frac{1+i\pi\mathcal{E}}{2}\right) \Gamma\left(\frac{1-i\pi\mathcal{E}}{2}\right)} h. \tag{11.47}$$

Proof. Using the the covariance of the calculus associated to the active symbol, it suffices to prove (11.45) in the case when f is the Dirac mass at the base point i of Π: then, from the definition (6.28) of the active symbol, the associated operator is $2\sigma^{\frac{1}{2}}$, as defined in (6.27):

$$(\sigma^{\frac{1}{2}}v)(s) = 2\pi \int_0^\infty v(t) \left(\frac{s}{t}\right)^{\frac{1}{4}} J_{\frac{1}{2}}\left(4\pi\sqrt{st}\right) dt. \tag{11.48}$$

Now, as computed in [26, p. 96], one has $\sigma^{\frac{1}{2}} = \exp i\pi \left(\mathrm{Lag} - \frac{3}{4}\right)$, where the Laguerre operator Lag has the discrete spectrum $\{\frac{3}{4} + k : k = 0, 1, \dots\}$. In other words, the operator $\sigma^{\frac{1}{2}}$ preserves the eigenfunctions of Lag with eigenvalues $\frac{3}{4}, \frac{11}{4}, \dots$ while changing those with eigenvalues $\frac{7}{4}, \frac{15}{4}, \dots$ to their negatives. The conjugate of $\sigma^{\frac{1}{2}}$ under the map $\mathrm{Sq}_{\mathrm{odd}}$, which of course kills all even functions, preserves the eigenfunctions of the harmonic oscillator $L = \pi \left(t^2 - \frac{1}{4\pi^2} \frac{d^2}{dt^2}\right)$ with eigenvalues $\frac{3}{2}, \frac{11}{2}, \dots$ while changing those with eigenvalues $\frac{7}{2}, \frac{15}{2}, \dots$ to their negatives. In other words, it coincides with the operator $\exp \frac{i\pi}{2}(L - \frac{3}{2})$.

As proved in [25, p. 204], the Weyl symbol of the operator $\exp(-\alpha L)$, for $|\mathrm{Im}\,\alpha| < \pi$, is the function

$$(q, p) \mapsto \left(\cosh \frac{\alpha}{2}\right)^{-1} e^{-2\pi(\tanh \frac{\alpha}{2})(q^2+p^2)}. \tag{11.49}$$

Finally, the Weyl symbol of the operator $\exp \frac{i\pi}{2}(L - \frac{3}{2})$ is $2^{\frac{1}{2}} e^{-\frac{3i\pi}{4}} e^{-2i\pi(q^2+p^2)}$, and it is easy to see that the $\mathcal{G}$-antiinvariant part of the symbol $e^{-2i\pi\|(q,p)\|_z^2}$ is the function $e^{\frac{i\pi}{4}} \cos\left(2\pi\|(q,p)\|_z^2 + \frac{\pi}{4}\right)$, which leads to (11.45).

In the reverse direction, we recall that the active and passive symbols of an operator on $H_{\frac{3}{2}}$ are dual species in the sense given in the introduction of Chapter 2, while the isometric horocyclic symbol is self-dual. It follows that the map $h^{\mathrm{iso}} \mapsto f^{\mathrm{pass}}$ is the adjoint of Θ, hence given by the equation

$$f^{\mathrm{pass}}(z) = 2^{\frac{3}{2}} i \int_{\mathbb{R}^2} h^{\mathrm{iso}}(q, p) \cos\left(2\pi\|(q,p)\|_z^2 + \frac{\pi}{4}\right) dq\,dp. \tag{11.50}$$

One then takes advantage of (6.31) to obtain f^{act} in terms of f^{pass}, using again the fact that, under an integral transform such as (11.50), the operator $\Delta - \frac{1}{4}$ is the transfer of $\pi^2 \mathcal{E}^2$. This concludes the proof of Lemma 11.3. $\quad\square$

Equation (11.45), together with (11.46), makes it possible to decompose any $\mathcal{G}$-antiinvariant function $h \in \mathcal{S}_{\mathrm{even}}(\mathbb{R}^2)$ as an integral superposition of $\mathcal{G}$-antiinvariant functions, each of which is radial relative to a certain Euclidean structure on $\mathbb{R}^2$: however, these functions are far from lying in $\mathcal{S}(\mathbb{R}^2)$. A cure (respecting the $\mathcal{G}$-antiinvariance) consists in replacing $h = \Theta f$ by the approximation (in which $\varepsilon > 0$)

$$h_\varepsilon(q,\,p) = (\Theta_\varepsilon f)(q,\,p)$$

$$= (1 - i) \int_\Pi f(z) \left[e^{-2\pi(\varepsilon - i)\,\|\,(q,p)\,\|_z^2} - \frac{1}{\varepsilon - i} e^{-\frac{2\pi}{\varepsilon - i}\,\|\,(q,p)\,\|_z^2} \right] d\mu(z).$$

$$(11.51)$$

We wish to be brief at that point: however, let us show to begin with that such a transformation indeed produces a function in $L^2(\mathbb{R}^2)$, provided one starts from a function $f \in L^2(\Pi)$ which can be written as $f = (\Delta - \frac{1}{4}) f_1$ with $f_1 \in L^2(\Pi)$. Denoting as h_ε^+ the first term of the decomposition of h_ε provided by (11.51), one has $\| h_\varepsilon^+ \|_{L^2(\mathbb{R}^2)}^2 = \int_{\Pi \times \Pi} K(w,\,z)\,\overline{f}(w)\,f(z)\,d\mu(w)\,d\mu(z)$ with

$$K(w,\,z) = \int_{\mathbb{R}^2} e^{-2\pi(\varepsilon + i)\,\|\,(q,p)\,\|_w^2}\, e^{-2\pi(\varepsilon - i)\,\|\,(q,p)\,\|_z^2}\, dq\,dp$$

$$= \int_{\mathbb{R}^2} e^{-2\pi\,Q(q,\,p)}\,dq\,dp = \frac{1}{2}\,(\det Q)^{-\frac{1}{2}}, \qquad (11.52)$$

where the determinant of the matrix representing the quadratic form Q is (setting $z = x + iy$, $w = u + iv$)

$$\det Q = 2\,(\varepsilon^2 - 1) + (1 + \varepsilon^2)\,\frac{(x - u)^2 + v^2 + y^2}{vy}$$

$$= 2\,\left[(1 + \varepsilon^2)\,\cosh d(z,\,w) + \varepsilon^2 - 1 \right]. \qquad (11.53)$$

Now, on functions of $\delta = \cosh d(z,\,w)$ regarded as functions of, say, z only, Δ expresses itself as $(1 - \delta^2)\,\frac{d^2}{d\delta^2} - 2\delta\,\frac{d}{d\delta}$ and, if $c < 1$, one obtains

$$\left(\Delta - \frac{1}{4} \right)\,(\delta - c)^{-\frac{1}{2}} = (\delta - c)^{-\frac{5}{2}}\,\left(-\frac{c\delta}{2} + \frac{3}{4} - \frac{c^2}{2} \right). \qquad (11.54)$$

Finally, this is a summable function with respect to $d\mu$, a measure which, on functions of $\cosh r = \cosh d(z,\,w)$, reduces to $4\pi \sinh r\,dr$. This justifies our claim made just after (11.51).

With the help of the equation [31, p. 17]

$$\Theta_\varepsilon \left(\Delta - \frac{1}{4} \right) f_1 = \pi^2 \mathcal{E}^2\,\Theta_\varepsilon f_1 \qquad (11.55)$$

and of (11.41), it is easy to see that the space

$$E = \text{linear span of } \left\{ h_\varepsilon = \Theta_\varepsilon \left(\Delta - \frac{1}{4} \right) f_1 \colon f_1 \in C_0^\infty(\Pi),\ \varepsilon > 0 \right\} \qquad (11.56)$$

satisfies the following "density" property: every even $\mathcal{G}$-antiinvariant function $h \in L^2(\mathbb{R}^2)$ can be approached by functions in E in the topology of $L_{\text{loc}}^2(\mathbb{R}^2)$.

Theorem 11.5. *Let $\tau > -1$ be given, let $\mathfrak{s}_\tau$ be a τ-adapted distribution in the sense of Definition 9.1, and let f be the holomorphic function (a modular form of real weight $\tau + 1$) associated to $\mathfrak{s}_\tau$ by means of (9.2). Define a Γ-automorphic distribution by the equation*

$$W^\tau(\mathfrak{s}_\tau, \mathfrak{s}_\tau) = \frac{2^{\tau-2}\pi^{\tau-\frac{1}{2}}}{\Gamma(\tau + \frac{3}{2})} \left[\int_{-\infty}^{\infty} \frac{L^*(\bar{f} \otimes f, \frac{1-i\lambda}{2})}{\zeta(i\lambda)\,\zeta(-i\lambda)} \, i\lambda \, \mathfrak{F}_{i\lambda} \, d\lambda \right.$$

$$\left. - 24\pi \, \| y^{\frac{\tau+1}{2}} f \|^2_{L^2(\Gamma\backslash\Pi)} \left(\frac{\delta}{2} - 1 \right). \right] \quad (11.57)$$

and note that it is $\mathcal{G}$-antiinvariant in view of (11.4) and (2.8). It is a soft Wigner function of the pair $(\mathfrak{s}_\tau, \mathfrak{s}_\tau)$ in the following weak sense: given any symbol h in the linear space E just defined, one has the identity (cf. (10.3))

$$\mathrm{Tr}\,(\mathrm{Op}^\tau_{\mathrm{soft}}(h)\, P_{\mathfrak{s}_\tau, \mathfrak{s}_\tau}) = \langle\, h\,,\, W^\tau(\mathfrak{s}_\tau, \mathfrak{s}_\tau)\,\rangle. \quad (11.58)$$

Proof. First, we prove that if one substitutes for h any $\mathcal{G}$-antiinvariant *radial* symbol in $\mathcal{S}(\mathbb{R}^2)$, also the image under $i\pi\mathcal{E}$ of a $\mathcal{G}$-invariant *radial* symbol in $\mathcal{S}(\mathbb{R}^2)$, one has the more general identity

$$\mathrm{Tr}\,(\mathrm{Op}^\tau_{\mathrm{soft}}(h)\, P_{\mathfrak{s}^g_\tau, \mathfrak{s}^g_\tau}) = \langle\, h\,,\, W^\tau(\mathfrak{s}_\tau, \mathfrak{s}_\tau) \circ g\,\rangle \quad (11.59)$$

for every $g \in G$. For $\mathrm{Re}\,\nu < -1$, let us compute, using (11.1) and (11.3),

$$\langle\, h\,,\, \mathfrak{F}_\nu\,\rangle = 2^{\frac{-3-\nu}{2}} \sum_{|m|+|n|\neq 0} \int_{-\infty}^{\infty} |t|^{-\nu}\, h(tn, tm)\, dt$$

$$= 2^{\frac{-3-\nu}{2}} \zeta(1-\nu) \sum_{(m,n)=1} \int_{-\infty}^{\infty} |t|^{-\nu}\, h(tn, tm)\, dt. \quad (11.60)$$

Using (5.11), one can write

$$\int_{-\infty}^{\infty} |t|^{-\nu}\, h(tn, tm)\, dt = 2 \int_0^{\infty} t^{-\nu}\, H(t^2(n^2 + m^2))\, dt$$

$$= 2\,(n^2 + m^2)^{\frac{\nu-1}{2}} \int_{-\infty}^{\infty} |t|^{-\nu}\, H(t^2)\, dt$$

$$= 4\pi\,(n^2 + m^2)^{\frac{\nu-1}{2}} \psi(\nu). \quad (11.61)$$

Hence,

$$\langle\, h\,,\, \mathfrak{F}_\nu\,\rangle = 2^{\frac{1-\nu}{2}} \pi\, \zeta(1-\nu)\, \psi(\nu) \sum_{(m,n)=1} (n^2 + m^2)^{\frac{\nu-1}{2}}$$

$$= 2^{\frac{3-\nu}{2}} \pi\, \zeta(1-\nu)\, \psi(\nu)\, E\left(i, \frac{1-\nu}{2} \right): \quad (11.62)$$

since, when $z = \frac{ai+b}{ci+d}$, one has $(c^2+d^2)\,m^2-2\,(ac+bd)\,mn+(a^2+b^2)\,n^2 = \frac{|m-nz|^2}{\mathrm{Im}\,z}$, one also has, using also analytic continuation,

$$\langle h \circ g^{-1}, \mathfrak{F}_{i\lambda} \rangle = 2^{\frac{3-i\lambda}{2}}\, \pi\, \zeta(1 - i\lambda)\, \psi(i\lambda)\, E\left(z, \frac{1 - i\lambda}{2}\right). \tag{11.63}$$

Using the equations (9.46)

$$L^*\left(\overline{f} \otimes f, \frac{1 - i\lambda}{2}\right) = \zeta^*(1 - i\lambda) \times \frac{\Gamma(\frac{1-i\lambda}{2} + \tau)}{(4\pi)^{\frac{1-i\lambda}{2}+\tau}}\, L\left(\overline{f} \otimes f, \frac{1 - i\lambda}{2}\right) \tag{11.64}$$

and

$$E(z, \frac{1 - i\lambda}{2}) = \frac{E^*(z, \frac{1-i\lambda}{2})}{\zeta^*(1 - i\lambda)}, \tag{11.65}$$

together with (9.53), one can write the right-hand side of (11.59) as

$$\frac{2^{\tau-2}\,\pi^{\tau-\frac{1}{2}}}{\Gamma(\tau + \frac{3}{2})} \int_{-\infty}^{\infty} \frac{\Gamma(\frac{1-i\lambda}{2} + \tau)}{(4\pi)^{\frac{1-i\lambda}{2}+\tau}}\, \frac{L(\overline{f} \otimes f, \frac{1-i\lambda}{2})}{\zeta(i\lambda)\,\zeta(-i\lambda)}$$

$$\times\, 2^{\frac{3-i\lambda}{2}}\, \pi\, \zeta(1 - i\lambda)\, \psi(i\lambda)\, i\lambda\, E^*\left(z, \frac{1 - i\lambda}{2}\right)\, d\lambda$$

$$-\, 2^{-\tau-1}\,\pi^{\frac{1}{2}}\, \frac{\Gamma(\tau + 1)}{\Gamma(\tau + \frac{3}{2})}\, h(0)\, \mathrm{Res}_{s=1}\, (L(\overline{f} \otimes f), s)) \tag{11.66}$$

an expression which can be identified with the left-hand side of (11.58), as given in (10.56), using the equation

$$\frac{\zeta(1 - i\lambda)}{\zeta(i\lambda)\,\zeta(-i\lambda)} = \pi^{\frac{1-i\lambda}{2}}\, \frac{\Gamma(\frac{i\lambda}{2})\,\Gamma(-\frac{i\lambda}{2})}{\Gamma(\frac{1-i\lambda}{2})\,\zeta^*(-i\lambda)}. \tag{11.67}$$

To prove (11.58) in the case when h is the transform of a radial symbol, with the same properties as before, under the linear change of coordinates associated with an arbitrary matrix $g \in G$, it suffices to use (11.59) together with the pair of equations

$$P_{s^g_\tau, s^g_\tau} = (\mathcal{D}_{\tau+1}(g))^{-1}\, P_{s_\tau, s_\tau}\, \mathcal{D}_{\tau+1}(g),$$
$$\mathrm{Op}_\tau(h \circ g^{-1}) = \mathcal{D}_{\tau+1}(g)\, \mathrm{Op}_\tau(h)\, (\mathcal{D}_{\tau+1}(g))^{-1}. \tag{11.68}$$

The general case follows, only using an integral superposition of equations just obtained. $\square$

Corollary 11.6. *Under the conditions of the preceding theorem, one can also give the operator $P_{\mathfrak{s}_\tau, \mathfrak{s}_\tau}$ an isometric horocyclic symbol $(W^\tau)^{\text{iso}}(\mathfrak{s}_\tau, \mathfrak{s}_\tau)$, by definition (cf. Definition 7.3) the image of $W^\tau(\mathfrak{s}_\tau, \mathfrak{s}_\tau)$ by the inverse of the operator $\frac{2^{-\tau}}{\Gamma(\tau+\frac{3}{2})} \Gamma(1 - i\pi\mathcal{E}) \Gamma(\tau + \frac{1}{2} + i\pi\mathcal{E})$. Its decomposition into homogeneous components is given by the equation*

$$(W^\tau)^{\text{iso}}(\mathfrak{s}_\tau, \mathfrak{s}_\tau) = \frac{1}{2\pi} \int_{-\infty}^{\infty} 2^{\frac{i\lambda-3}{2}} \frac{L(\overline{f} \otimes f, \frac{1-i\lambda}{2})}{\zeta(-i\lambda)} \mathfrak{E}_{i\lambda} \, d\lambda$$
$$+ \frac{1}{2} \operatorname{Res}_{s=1} \left(L(\overline{f} \otimes f, s) \right) (1 - \tau \delta). \quad (11.69)$$

Proof. Since $2i\pi\mathcal{E}$ acts on $\mathfrak{E}_{i\lambda}$ like the multiplication by $-i\lambda$, and on δ *(resp. 1)* like the multiplication by -1 *(resp. 1)*, the corollary is a consequence of simple calculations using again (9.46) and (9.53). $\qquad \square$

This provides a means of verification. In the case when $\tau = -\frac{1}{2}$ and $\mathfrak{s}_\tau$ is the τ-adapted distribution associated to the modular form $f^{(12)}$ of Section 3, the formula can be written, with the help of (9.47), as

$$\left(W^{-\frac{1}{2}}\right)^{\text{iso}}(\mathfrak{s}_{-\frac{1}{2}}, \mathfrak{s}_{-\frac{1}{2}}) = \frac{1}{2\pi} \int_{-\infty}^{\infty} 2^{\frac{i\lambda-3}{2}} 24^{-\frac{i\lambda}{2}} (1 - 2^{i\lambda})(1 - 3^{i\lambda}) \mathfrak{E}_{i\lambda} \, d\lambda$$
$$+ 2^{-\frac{1}{2}} 3^{-\frac{1}{2}} \left(1 + \frac{\delta}{2}\right). \quad (11.70)$$

When $\tau = \frac{1}{2}$ and $\mathfrak{s}_\tau$ is the τ-adapted distribution associated to the modular form $f^{(4)}$, it can be written, with the help of (9.48), as

$$\left(W^{\frac{1}{2}}\right)^{\text{iso}}(\mathfrak{s}_{\frac{1}{2}}, \mathfrak{s}_{\frac{1}{2}}) = \frac{1}{2\pi} \int_{-\infty}^{\infty} 2^{-\frac{1}{2}} (2^{-i\lambda} - 1) \mathfrak{E}_{i\lambda} \, d\lambda + 2^{-\frac{1}{2}} \left(1 - \frac{\delta}{2}\right). \quad (11.71)$$

One can then compare the formulas obtained with the last two equations (4.36). If one uses the decomposition (11.5) of $\mathfrak{D}_0$ into homogeneous components, one finds that

$$W(\mathfrak{d}_{\text{even}}, \mathfrak{d}_{\text{even}}) = 2^{\frac{3}{2}} \left(W^{-\frac{1}{2}}\right)^{\text{iso}}(\mathfrak{s}_{-\frac{1}{2}}, \mathfrak{s}_{-\frac{1}{2}}),$$
$$W(\mathfrak{d}_{\text{odd}}, \mathfrak{d}_{\text{odd}}) = 2^{\frac{1}{2}} \left(W^{\frac{1}{2}}\right)^{\text{iso}}(\mathfrak{s}_{\frac{1}{2}}, \mathfrak{s}_{\frac{1}{2}}) : \quad (11.72)$$

Remark 9.1, towards the end of Section 9, explains the origin of the extra factors $2^{\frac{3}{2}}$ and $2^{\frac{1}{2}}$.

Remark 11.1. Corollary 11.6 does not interpret zeros of the function $\lambda \mapsto L(\overline{f} \otimes f, \frac{1-i\lambda}{2})$ as eigenvalues of some operator, rather as zeros of some spectral density. Such a phenomenon is of course far from unusual when dealing with functions such

as zeta or L-functions. The simplest example is probably provided by the formula
[32, p. 1158]

$$\eth_0(x) = 1 + \delta + \frac{1}{2\pi} \int_{-\infty}^{\infty} \zeta\left(\frac{1}{2} - i\lambda\right) |x|^{-\frac{1}{2} - i\lambda} \, d\lambda \qquad (11.73)$$

which, in the same way as Proposition 3.3, gives the decomposition into homogeneous components of the Dirac comb. In contrast, Corollary 11.6 deals with a whole class of distributions: as already asked in Section 9, a possibly interesting question seems to be the dependence on τ of the collection of zeros in question, when f is chosen, say, as in (9.3).

Chapter 4

Back to the Weyl Calculus

We come back to the constructions of Section 3: our standing assumption in this chapter is that the number N is 4 times a squarefree odd integer, but it does not have to coincide with 4 or 12 any more. Then, not every element of $(\mathbb{Z}/N\mathbb{Z})^{\times}$ is a square, and we have to consider the full set of distributions ϖ_ρ as defined in Lemma 3.2: we introduce the linear combination

$$W_N(x, \xi) = \sum_{\rho \in R_N} W(\varpi_\rho, \varpi_\rho)(x, \xi),$$

again a Γ-invariant distribution. Recall that Λ is the set of squares in $(\mathbb{Z}/N\mathbb{Z})^{\times}$ and that R_N is a set of representatives of $(\mathbb{Z}/N\mathbb{Z})^{\times}$ mod Λ.

The main result of Section 12 is that a certain distribution $\mathfrak{T}_\infty$ is a limit, as $N \to \infty$, of a distribution $\mathfrak{T}_N$ closely related to W_N. The point is that the most obvious question regarding the decomposition of $\mathfrak{T}_\infty$ into homogeneous components is equivalent to the Riemann hypothesis. This is certainly not to mean that the present investigations provide a new approach since, as explained at the end of Section 12, the occurrence of the N-dependent unitary operator $N^{-i\pi\mathcal{E}}$ prevents giving the limit $\mathfrak{T}_\infty$ an interpretation as a Wigner function. In the last section, we bring forward the role of combs or, equivalently, Dirichlet series, and make some steps in the direction of building an analysis specially adapted to this kind of objects.

12 Letting N go to infinity

Theorem 12.1. *With the notation of Section 3 and that in (4.35), one has*

$$W_N = N^{i\pi\mathcal{E}} \prod_{p \in S} (1 - p^{-2i\pi\mathcal{E}}) \mathfrak{D}_0. \tag{12.1}$$

Proof. Because of the bijection between R_N and $\prod_{p \in S} R_p$, as mentioned just after (4.26), this computation (starting from (4.19)) can be localized too, ending up with

$$W_N(x, \xi) = \frac{1}{\sqrt{N}} \sum_{j,k \in \mathbb{Z}} \Gamma_{jk}^{(N)} \, \delta(x - \frac{j}{\sqrt{N}}) \, \delta\left(\xi - \frac{k}{\sqrt{N}}\right), \tag{12.2}$$

where

$$\Gamma_{jk}^{(N)} = \prod_{p \in S} \Gamma_{jk}^{(p)}, \qquad \Gamma_{jk}^{(p)} := \sum_{\rho \in R_p} \Gamma_{jk}^{(p)}(\rho, \rho). \tag{12.3}$$

When $p = 2$, we have already obtained, in (4.33), that $\Gamma_{jk}^{(2)} = -1$ if j and k are even, 1 in all other cases. We now assume $p \geq 3$, so that $\alpha_p = 1$. Then, $\Lambda_p = \{\pm 1\} \subset \mathbb{Z}/p\mathbb{Z}$, and

$$\Gamma_{jk}^{(p)}(\rho, \rho) = \sum_{\substack{\mu_1, \mu_2 \in \Lambda_p \\ \rho(\mu_1 + \mu_2) \equiv 2j \bmod p}} \chi_p(\mu_1) \, \chi_p(\mu_2) \, \exp\left(\frac{2i\pi \overline{M} k}{p} \rho(\mu_1 - \mu_2)\right). \tag{12.4}$$

First, assume that $j \not\equiv 0 \bmod p$. Then, there is at most one non-zero term in the sum: it occurs if and only if $\rho \equiv \pm j$ and has value 1. When $j \equiv 0$, the terms with $\mu_1 = \mu_2$ contribute the value 2 to the sum if $\rho \equiv 0$, and do not contribute to the sum if $\rho \not\equiv 0$. On the other hand, the terms with $(\mu_1, \mu_2) = (1, -1)$ and $(-1, 1)$ together contribute the value $-2 \cos \frac{4\pi \overline{M} k\rho}{p}$. Hence, not forgetting that, in the first case, only one of the two classes $\pm j \bmod p$ lies in R_p, a set of representatives mod $\{\pm 1\}$,

$$\Gamma_{jk}^{(p)} = \begin{cases} 1 & \text{if } j \not\equiv 0 \bmod p, \\ -2 \sum_{\rho \in R_p} \cos \frac{4\pi \overline{M} k\rho}{p} & \text{if } j \equiv 0 \bmod p. \end{cases} \tag{12.5}$$

Now,

$$\sum_{\rho \in R_p} \cos \frac{4\pi \overline{M} k\rho}{p} = \frac{1}{2} \left[-1 + \sum_{\rho \in \mathbb{Z}/p\mathbb{Z}} \cos \frac{4\pi \overline{M} k\rho}{p} \right]$$

$$= \frac{1}{2} \left[-1 + \sum_{\rho \in \mathbb{Z}/p\mathbb{Z}} \exp\left(\frac{2i\pi \overline{M} k\rho}{p}\right) \right]$$

$$= \begin{cases} -\frac{1}{2} & \text{if } k \not\equiv 0 \bmod p, \\ \frac{p-1}{2} & \text{if } k \equiv 0 \bmod p, \end{cases} \tag{12.6}$$

so that

$$\Gamma_{00}^{(p)} = 1 - p, \qquad \Gamma_{jk}^{(p)} = 1 \text{ if } (j,k) \neq (0,0). \tag{12.7}$$

Recall that this is also true if $p = 2$ (we have already obtained, too, the result when $p = 3$ in (4.34)), so that

$$\Gamma_{jk}^{(N)} = \prod_{\substack{p \in S \\ p|j,\, p|k}} (1 - p).\tag{12.8}$$

On the other hand, let us compute

$$\dot{W}_N := N^{i\pi\mathcal{E}} \prod_{p \in S} (1 - p^{-2i\pi\mathcal{E}})\, \mathfrak{D}_0.\tag{12.9}$$

Introduce the set of all vectors $\epsilon = (\varepsilon_p)_{p \in S}$ such that $\varepsilon_p = 0$ or 1 for every $p \in S$: then, setting $|\epsilon| = \sum \varepsilon_p$, one finds

$$\dot{W}_N = N^{i\pi\mathcal{E}} \sum_{\epsilon} (-1)^{|\epsilon|} \left(\prod_p p^{-2i\pi\varepsilon_p\mathcal{E}} \right) \mathfrak{D}_0$$

$$= \sum_{\epsilon} (-1)^{|\epsilon|} \left(N^{-\frac{1}{2}} \prod_p p^{\varepsilon_p} \right)^{-2i\pi\mathcal{E}} \mathfrak{D}_0,\tag{12.10}$$

or, using (4.23),

$$\dot{W}_N(x, \xi) = \frac{1}{\sqrt{N}} \sum_{m,n \in \mathbb{Z}} \sum_{\epsilon} (-1)^{|\epsilon|} \prod_p p^{\varepsilon_p}$$

$$\delta \left(x - N^{-\frac{1}{2}} \prod_p p^{\varepsilon_p}\, m \right) \delta \left(\xi - N^{-\frac{1}{2}} \prod_p p^{\varepsilon_p}\, n \right).\tag{12.11}$$

For any given pair $(j, k) \in \mathbb{Z} \times \mathbb{Z}$, the coefficient of $\delta(x - \frac{j}{\sqrt{N}})\, \delta(\xi - \frac{k}{\sqrt{N}})$ in this sum is the sum

$$\frac{1}{\sqrt{N}} \sum \left\{ (-1)^{|\epsilon|} \prod_p p^{\varepsilon_p} : \epsilon \text{ such that } \prod_p p^{\varepsilon_p} | j \text{ and } \prod_p p^{\varepsilon_p} | k \right\}.\tag{12.12}$$

In view of (12.8), this is the same as the coefficient of $\delta(x - \frac{j}{\sqrt{N}})\, \delta(\xi - \frac{k}{\sqrt{N}})$ in (12.2), which concludes the proof. $\qquad\square$

The next lemma, to the effect that a certain automorphic function is rapidly decreasing at infinity in the fundamental domain, relies on the fact that the discretely supported distributions ϖ_ρ have no mass at the origin.

Lemma 12.2. *Let S be a finite set of primes including 2, and let $N = 2 \prod_{p \in S} p$. With the notation of Lemma 3.2, and recalling that the functions u_z and u_z^1 have been defined in (2.12), the function*

$$\phi(z) = \begin{cases} \sum_{\rho \in R_N} |\langle \varpi_\rho,\, u_z \rangle|^2 & \text{if } \#S \text{ is even}, \\ \sum_{\rho \in R_N} |\langle \varpi_\rho,\, u_z^1 \rangle|^2 & \text{if } \#S \text{ is odd} \end{cases}\tag{12.13}$$

is automorphic for the full modular group (i.e., invariant under linear transfor-
mations associated to matrices in $SL(2,\mathbb{Z})$*) and rapidly decreasing at infinity in*
the fundamental domain.

Proof. From their definition in Lemma 3.2, the distributions ϖ_ρ are even or odd
according to the parity of $\#S$, the number of distinct primes, including 2, dividing
N. We shall consider the even case only, since the other one is entirely similar.
Using the basic property (2.4) of Wigner functions, together with the identity
$W_N = \dot{W}_N$ from the proof of Theorem 12.1, finally (12.2) and (2.20), one obtains

$$\phi(z) = \frac{2}{\sqrt{N}} \sum_{j,k\in\mathbb{Z}} \Gamma_{jk}^{(N)} \exp\left(-\frac{2\pi}{Ny}\,|j-kz|^2\right) \tag{12.14}$$

with $\Gamma_{jk}^{(N)}$ as made explicit in (12.8). That ϕ is automorphic is obvious. Let us
first consider the sum of terms with $k=0$, *i.e.*,

$$\psi(y) = \sum_{j\in\mathbb{Z}} \left(\prod_{p\in S,\,p|j} (1-p)\right) e^{-\frac{2\pi j^2}{Ny}}. \tag{12.15}$$

Let A be a set of representatives mod $\frac{N}{2}$ of the set of odd numbers: then, the
disjoint sum $A\cup B$, where $B = \{2m\colon m\in A\}$, is a set of representatives of $\mathbb{Z}$
mod $\frac{N}{2}$. If $m\in A$, one has

$$\prod_{p\in S,\,p|2m} (1-p) = -\prod_{p\in S,\,p|m} (1-p), \tag{12.16}$$

so that $\psi = \sum_{m\in A}\psi_m$ with

$$\psi_m(y) = \prod_{p\in S,\,p|m} (1-p) \tag{12.17}$$

$$\times\left[\sum_{\ell\in\mathbb{Z}} \exp\left(-\frac{2\pi}{Ny}\left(m+\frac{\ell N}{2}\right)^2\right) - \sum_{\ell\in\mathbb{Z}} \exp\left(-\frac{2\pi}{Ny}\left(2m+\frac{\ell N}{2}\right)^2\right)\right].$$

Note that the equation just obtained, making such a difference appear, is the
important part of the argument: nothing analogous could work if, say, $\mathfrak{D}_1$, as
introduced in Theorem 4.2, were substituted for W_N. Indeed, the function $z\mapsto$
$(\mathfrak{D}_0, u_z)$ is not square-integrable in the fundamental domain of the group Γ_2 of
invariance of $\mathfrak{D}_0$. Set

$$f_m(x) = \exp\left(-\frac{2\pi}{Ny}\left(m+\frac{Nx}{2}\right)^2\right)$$

$$= h\left(y^{-\frac{1}{2}}\left(x+\frac{2m}{N}\right)\right) \tag{12.18}$$

with $h(x) = e^{-\frac{\pi N x^2}{2}}$. Then, for every $n = 0, 1, \ldots$,

$$f^{(n)}(x) = y^{-\frac{n}{2}} h^{(n)} \left(y^{-\frac{1}{2}} \left(x + \frac{2m}{N} \right) \right), \qquad (12.19)$$

so that

$$\int_{-\infty}^{\infty} |f^{(n)}(x)| \, dx \le C \, y^{\frac{1-n}{2}} \qquad (12.20)$$

for some constant $C > 0$ depending only on N. It then follows from the Euler-Maclaurin formula (*cf.* e.g. [5, p. 8]) that each of the two terms in the difference on the right-hand side of (12.17) can be written as

$$\left(\frac{2y}{N} \right)^{\frac{1}{2}} + O \left(y^{\frac{1-n}{2}} \right). \qquad (12.21)$$

This proves that $\psi(y)$ goes to zero rapidly as $y \to \infty$.

That the terms with $k \ne 0$ contribute to a sum which goes rapidly to zero as $y \to \infty$ is obvious. $\qquad \square$

Given any radial symbol $h \in \mathcal{S}(\mathbb{R}^2)$, we set

$$(\mathcal{A}h)^{(N)}(z) = \sum_{\rho \in R_N} (\mathrm{Met}(\tilde{g}^{-1}) \, \varpi_\rho \, | \, \mathrm{Op}(h) \, \mathrm{Met}(\tilde{g}^{-1}) \, \varpi_\rho), \qquad (12.22)$$

where $z = g.i \in \Pi$ and $\tilde{g}$ is a point of the metaplectic group lying above g. In particular, $(\mathcal{A}h)^{(12)} = (\mathcal{A}h)_0$ and $(\mathcal{A}h)^{(4)} = (\mathcal{A}h)_1$.

Theorem 12.3. *Let $h \in \mathcal{S}(\mathbb{R}^2)$ satisfy moreover the condition that $h = \mathcal{G}h$ if $\#S$ is even, and $h = -\mathcal{G}h$ if $\#S$ is odd. Set*

$$\zeta_N(s) = \prod_{p \in S} (1 - p^{-s})^{-1}. \qquad (12.23)$$

Then, one has

$$(\mathcal{A}h)^{(N)}(z) = 2 \int_{-\infty}^{\infty} \frac{N^{-\frac{i\lambda}{2}}}{\zeta_N(-i\lambda)} \frac{\pi^{\frac{1-i\lambda}{2}}}{\Gamma(\frac{1-i\lambda}{2})} \, \psi(i\lambda) \, E^* \left(z, \frac{1-i\lambda}{2} \right) d\lambda$$

$$+ (-1)^{\#S} \frac{\phi_\mathcal{E}(N)}{N^{\frac{1}{2}}} h(0) : \quad (12.24)$$

recall that $\phi_\mathcal{E}$ is Euler's indicator function. If, moreover, $h(0) = 0$, one has the identity

$$\| (\mathcal{A}h)^{(N)} \|^2_{L^2(\Gamma \backslash \Pi)} = \frac{8}{\pi} \left\| \frac{\zeta(1 - 2i\pi\mathcal{E})}{\zeta_N(2i\pi\mathcal{E})} h \right\|^2_{L^2(\mathbb{R}^2)}. \qquad (12.25)$$

Proof. The proof of the second part is entirely similar to that of Corollary 5.2, making use of the expression (12.1) of the linear combination W_N of Wigner functions. For the first part, we must follow the proof of Theorem 5.1, setting

$$h^{(N)} = N^{-i\pi\mathcal{E}} \prod_{p\in S}(1 - p^{2i\pi\mathcal{E}})\, h \tag{12.26}$$

and $h^{(N)}(t, \theta) = H^{(N)}(t^2 + \theta^2)$ (one of the inconveniences of arithmetic quantization has to do with notation: we cannot use either (x, ξ) or (q, p) as a pair of variables on the phase space). Then, using the same notation as in the proof of Theorem 12.1,

$$H^{(N)}(\rho) = \sum_{\epsilon}(-1)^{|\epsilon|}\, N^{-\frac{1}{2}} \prod_{p\in S} p^{\varepsilon_p} \times H\left(N^{-1}\prod_{p\in S}p^{2\varepsilon_p}\,\rho\right), \tag{12.27}$$

so that

$$H^{(N)}(0) = N^{-\frac{1}{2}} \prod_{p\in S}(1 - p) \times h(0). \tag{12.28}$$

Then, (5.23) becomes

$$(\mathcal{A}h)^{(N)}(z) = N^{-\frac{1}{2}} \prod_{p\in S}(1 - p) \times h(0)$$
$$+ \frac{4}{i}\int_{\sigma-i\infty}^{\sigma+i\infty} N^{s-\frac{1}{2}} \prod_{p\in S}(1 - p^{1-2s}) \times \zeta(2s)\,\psi(1 - 2s)\, E(z, s)\, ds : \tag{12.29}$$

the pole at $s = 1$ contributes the term $\frac{1}{2}N^{\frac{1}{2}}\prod_{p\in S}(1 - p^{-1}) \times (\mathcal{G}h)(0)$: regrouping it with the first term on the right-hand side of (12.29), one easily obtains (12.24), using the elementary relation

$$\phi_{\mathcal{E}}(N) = N \prod_{p\in S}\left(1 - \frac{1}{p}\right) = 2\,(-1)^{\#S} \prod_{p\in S}(1 - p). \tag{12.30}$$

$\square$

The following theorem goes in the same direction as Theorem 9.3: its proof is quite different, except for the last part.

Theorem 12.4. *Let N be 4 times a product of distinct odd primes, and let us consider again the set R_N and the distributions ϖ_ρ, $\rho \in R_N$, introduced in Lemma 3.2: recall that S is the set of prime divisors of N. For every $\tilde{g}$ in the metaplectic group, lying above $g \in SL(2, \mathbb{R})$, set $\varpi_\rho^{\tilde{g}} = \mathrm{Met}(\tilde{g}^{-1})\,\varpi_\rho$, and let $u \in \mathcal{S}(\mathbb{R})$ be a function with the same parity as the number $\#S$. Denoting (abusively) as*

$|\langle \varpi_\rho^g, u \rangle|$ the absolute value $|\langle \varpi_{\bar\rho}^g, u \rangle|$, the sum $\sum_{\rho \in R_N} |\langle \varpi_\rho^g, u \rangle|^2$ only depends on the class Γg. Moreover, one has

$$\sum_{\rho \in R_N} \int_{\Gamma \backslash G} |\langle \varpi_\rho^g, u \rangle|^2 \, dg = \frac{2^{\frac{3}{2}} \pi}{3} \prod_{p \in S} (p^{\frac{1}{2}} - p^{-\frac{1}{2}}) \, \| u \|_{L^2(\mathbb{R})}^2$$

$$= \frac{2\pi}{3} N^{-\frac{1}{2}} \phi_\mathcal{E}(N) \, \| u \|_{L^2(\mathbb{R})}^2, \qquad (12.31)$$

where $\phi_\mathcal{E}$ is Euler's indicator function.

Proof. For the first point, it suffices, since $SL(2, \mathbb{Z})$ is generated by the matrices $\left(\begin{smallmatrix} 1 & 0 \\ 1 & 1 \end{smallmatrix} \right)$ and $\left(\begin{smallmatrix} 0 & 1 \\ -1 & 0 \end{smallmatrix} \right)$, to recall from what was said immediately after (3.18) that the matrices T and K representing, in the linear basis made up by the distributions ϖ_ρ, the multiplication by $e^{i\pi x^2}$ and the Fourier transformation are both unitary. Next, set $\epsilon = 0$ or 1 according to the parity of $\#S$. According to Lemma 12.2, the function

$$\phi(z) = \sum_{\rho \in R_N} |\langle \varpi_\rho, u_z^\epsilon \rangle|^2 \qquad (12.32)$$

is Γ-automorphic and rapidly decreasing at infinity. From that lemma and Theorem 12.1, this function can also be written as $\langle W_N, W(u_z^\epsilon, u_z^\epsilon) \rangle$ with (12.9)

$$W_N = N^{i\pi\mathcal{E}} \prod_{p \in S} (1 - p^{-2i\pi\mathcal{E}}) \, \mathfrak{D}_0. \qquad (12.33)$$

Starting from (11.5) and noting that, on the distributions $\delta, 1$ and $\mathfrak{E}_{i\lambda}$, the operator $2i\pi\mathcal{E}$ acts as the multiplication by $-1, 1$ and $-i\lambda$, one obtains

$$W_N = N^{-\frac{1}{2}} \prod_{p \in S} (1 - p) \, \delta + N^{\frac{1}{2}} \prod_{p \in S} (1 - p^{-1})$$

$$+ \frac{1}{2\pi} \int_{-\infty}^{\infty} N^{-\frac{i\lambda}{2}} \prod_{p \in S} (1 - p^{i\lambda}) \, \mathfrak{E}_{i\lambda} \, d\lambda. \qquad (12.34)$$

Let us go back to the role of Wigner functions, as explained in (2.4). Since the constant 1 is the (Weyl) symbol of the identity operator, and the functions u_z^ϵ are normalized, one has

$$\langle 1, W(u_z^\epsilon, u_z^\epsilon) \rangle = 1. \qquad (12.35)$$

Next, one has $\mathcal{G} 1 = \frac{1}{2} \delta$ and, looking back at the role, as explained in (2.9), of the transformation $\mathcal{G}$ on symbols, one obtains

$$\langle \delta, W(u_z^\epsilon, u_z^\epsilon) \rangle = 2 \langle 1, \mathcal{G} W(u_z^\epsilon, u_z^\epsilon) \rangle$$

$$= 2 (-1)^\epsilon \langle 1, W(u_z^\epsilon, u_z^\epsilon) \rangle$$

$$= 2 (-1)^\epsilon. \qquad (12.36)$$

Finally, using (11.35), (11.36) and (11.2), one has

$$\langle \mathfrak{E}_{i\lambda}, W(u_z, u_z) \rangle = 2^{\frac{1+i\lambda}{2}} E^*\left(z, \frac{1-i\lambda}{2}\right),$$

$$\langle \mathfrak{E}_{i\lambda}, W(u_z^1, u_z^1) \rangle = -2^{\frac{1+i\lambda}{2}} i\lambda\, E^*\left(z, \frac{1-i\lambda}{2}\right). \tag{12.37}$$

It follows that the constant term a in the expansion

$$\langle W_N, W(u_z^\epsilon, u_z^\epsilon) \rangle = a + \frac{1}{2\pi} \int_{-\infty}^{\infty} 2^{\frac{1}{2}} \prod_{p\in S}(p^{-\frac{i\lambda}{2}} - p^{\frac{i\lambda}{2}})\,(-i\lambda)^\epsilon\, E^*\left(z, \frac{1-i\lambda}{2}\right) d\lambda,$$

$$\tag{12.38}$$

which is also

$$a = \frac{3}{\pi} \int_{\Gamma\backslash\Pi} \phi(z)\, d\mu(z) \tag{12.39}$$

according to the general expansion theorem for automorphic functions in $L^2(\Gamma\backslash\Pi)$, is given as

$$a = N^{-\frac{1}{2}} \prod_{p\in S}(1-p) \times 2\,(-1)^\epsilon + N^{\frac{1}{2}} \prod_{p\in S}(1-p^{-1}) : \tag{12.40}$$

since $N = 2\prod_{p\in S} p$, this is also

$$a = 2^{\frac{3}{2}} \prod_{p\in S}(p^{\frac{1}{2}} - p^{-\frac{1}{2}}). \tag{12.41}$$

It follows that equation (12.31) is correct in the case when $u = u_z^\epsilon$, using also, for the second expression of the constant on the right-hand side, the elementary fact that $\phi_\mathcal{E}(N) = N \prod_{p\in S}(1 - \frac{1}{p})$.

Since u_z^ϵ is (cf. (2.12)) the product of some fixed power of Im $(-\frac{1}{z})$ by an antiholomorphic function of z, a "sesquiholomorphic" argument, similar to the one between (9.32) and the end of the proof of Theorem 9.3, makes it possible to conclude in general. $\square$

Since, as N increases to infinity, the union, as $\rho \in R_N$, of the supports of the distributions ϖ_ρ has a tendency towards filling up the whole line (in the right-hand side of (3.9), $\rho\mu$ can be any class in $(\mathbb{Z}/N\mathbb{Z})^\times$), one may expect that, up to some N-dependent normalisation, the operator $u \mapsto \sum_{\rho\in R_N}\langle \varpi_\rho, u \rangle\, \varpi_\rho$, or its Weyl symbol $W_N = \dot{W}_N$, may have a limit in some sense, possibly the identity operator. This is indeed the case: moreover, the renormalized limit of the error term will turn out to have a quite interesting spectral structure.

We shall assume that N goes to infinity in such a way that every finite set of primes should lie in S from a certain point on: we shall write $S \to P$, P denoting the set of all primes, to recall the fact.

Theorem 12.5. *Set*

$$W_N^\times = W_N - N^{-\frac{1}{2}} \prod_{p \in S} (1-p) \times \delta$$

$$= N^{-\frac{1}{2}} \sum_{\substack{|j|+|k| \neq 0}} \prod_{\substack{p \in S \\ p|j, p|k}} (1-p) \times \delta\left(x - \frac{j}{\sqrt{N}}\right) \delta\left(\xi - \frac{k}{\sqrt{N}}\right), \qquad (12.42)$$

and

$$\mathfrak{T}_N = N^{-i\pi\mathcal{E}} W_N^\times. \qquad (12.43)$$

As $N \to \infty$ in such a way that $S \to P$, the distribution $\mathfrak{T}_N$ converges, in the space of tempered automorphic distributions, towards the distribution $\mathfrak{T}_\infty$ such that

$$\mathfrak{T}_\infty(x, \xi) = \sum_{|j|+|k| \neq 0} \Gamma_{jk}^{(\infty)} \, \delta(x-j) \, \delta(\xi-k), \qquad (12.44)$$

where

$$\Gamma_{jk}^{(\infty)} = \prod_{\substack{p \text{ prime} \\ p|j, p|k}} (1-p). \qquad (12.45)$$

The decomposition into homogeneous components of the distribution $\mathfrak{T}_\infty$ is given as

$$\mathfrak{T}_\infty = \frac{1}{2\pi} \int_{-\infty}^{\infty} (\zeta(-i\lambda))^{-1} \, \mathfrak{E}_{i\lambda} \, d\lambda + \sum_{\zeta^*(\beta)=0} \mathrm{Res}_{\mu=\beta} \left(\frac{\mathfrak{E}_{-\mu}}{\zeta(\mu)} \right). \qquad (12.46)$$

Proof. The identity between the right-hand sides of the two equations (12.42) follows from (12.2) and (12.8). Hence, applying (4.23),

$$\mathfrak{T}_N(x, \xi) = \sum_{j,k \in \mathbb{Z}} \prod_{\substack{p \in S \\ p|j, p|k}} (1-p) \times \delta(x-j) \, \delta(\xi-k). \qquad (12.47)$$

Next, for any integer $d \geq 1$, denote as $\mathrm{M\ddot{o}b}(d)$ the value at d of the Möbius indicator function, *i.e.*, 0 if d is divisible by some square > 1, and ± 1 according to the parity of the number of prime divisors of k if this number is squarefree. Set

$$A(r) = \prod_{\substack{p \text{ prime} \\ p|r}} (1-p), \qquad (12.48)$$

so that

$$\Gamma_{jk}^{(\infty)} = A((j, k)). \qquad (12.49)$$

One has [14, p. 13]

$$A(r) = \sum_{1 \leq d|r} d \, \mathrm{M\ddot{o}b}(d), \qquad (12.50)$$

from which it follows that, provided Re $s > 2$,

$$\sum_{r \geq 1} r^{-s} A(r) = \sum_{r \geq 1} r^{-s} \sum_{1 \leq d | r} d \operatorname{M\ddot{o}b}(d)$$

$$= \sum_{d \geq 1} d \operatorname{M\ddot{o}b}(d) \sum_{m \geq 1} (md)^{-s}$$

$$= \zeta(s) \sum_{d \geq 1} d^{1-s} \operatorname{M\ddot{o}b}(d)$$

$$= \frac{\zeta(s)}{\zeta(s-1)}. \tag{12.51}$$

Let $h \in \mathcal{S}(\mathbb{R}^2)$. Coming back to the decomposition of functions in the plane into homogeneous components and generalizing (7.9), one can define

$$h_{-\mu}(x, \xi) = \frac{1}{2\pi} \int_0^\infty t^{-i\mu} h(tx, t\xi) \, dt \tag{12.52}$$

for every complex μ with Im $\mu > -1$, and one then has for every real number $a > 1$ (a condition to be taken advantage of later) the equation (using (7.8) and a contour deformation)

$$\langle \mathfrak{T}_\infty, h \rangle = \int_{ia-\infty}^{ia+\infty} \langle \mathfrak{T}_\infty, h_{-\mu} \rangle \, d\mu. \tag{12.53}$$

Then,

$$\langle \mathfrak{T}_\infty, h_{-\mu} \rangle = \frac{1}{2\pi} \sum_{|j|+|k| \neq 0} A((j, k)) \int_0^\infty t^{-i\mu} h(jt, kt) \, dt$$

$$= \frac{1}{2\pi} \sum_{r \geq 1} \sum_{(j, k)=1} A(r) \int_0^\infty t^{-i\mu} h(rjt, rkt) \, dt$$

$$= \frac{1}{2\pi} \sum_{r \geq 1} A(r) \, r^{-1+i\mu} \sum_{(j, k)=1} \int_0^\infty t^{-i\mu} h(jt, kt) \, dt$$

$$= \frac{1}{2\pi} \frac{\zeta(1-i\mu)}{\zeta(-i\mu)} \sum_{(j, k)=1} \int_0^\infty t^{-i\mu} h(jt, kt) \, dt$$

$$= \frac{1}{2\pi} \frac{\zeta(1-i\mu)}{\zeta(-i\mu)} \sum_{(j, k)=1} h_{-\mu}(j, k). \tag{12.54}$$

If one introduces [31, p. 23] the distribution $\mathfrak{D}^{\text{prime}}$ such that

$$\langle \mathfrak{D}^{\text{prime}}, h \rangle = 2\pi \sum_{(j, k)=1} h(j, k), \tag{12.55}$$

one thus has

$$\langle \mathfrak{T}_\infty, h_{-\mu} \rangle = \frac{1}{2\pi} \frac{\zeta(1 - i\mu)}{\zeta(-i\mu)} \langle \mathfrak{D}^{\text{prime}}, h_{-\mu} \rangle . \tag{12.56}$$

On the other hand, from (4.35), one has

$$\mathfrak{D}_0 = \delta + \frac{1}{2\pi} \mathfrak{D} \tag{12.57}$$

with

$$\mathfrak{D}(x, \xi) = 2\pi \sum_{|m|+|n| \neq 0} \delta(x - n) \delta(x - m) . \tag{12.58}$$

Also [30, p. 170],

$$\zeta(1 - i\mu) \langle \mathfrak{D}^{\text{prime}}, h_{-\mu} \rangle = \langle \mathfrak{D}, h_{-\mu} \rangle . \tag{12.59}$$

Finally [30, p. 164]

$$\langle \mathfrak{D}, h_{-\mu} \rangle = \langle \mathfrak{E}_{i\mu} , h \rangle \tag{12.60}$$

(note that, in the given reference, $\mathfrak{E}_{i\mu}$ was denoted $\mathfrak{E}_{i\mu}^{\sharp}$), so that

$$\langle \mathfrak{T}_\infty, h_{-\mu} \rangle = \frac{1}{2\pi} (\zeta(-i\mu))^{-1} \langle \mathfrak{E}_{i\mu} , h \rangle \tag{12.61}$$

and, from (12.53),

$$\begin{aligned}
\mathfrak{T}_\infty &= \frac{1}{2\pi} \int_{ia-\infty}^{ia+\infty} (\zeta(-i\mu))^{-1} \mathfrak{E}_{i\mu} \, d\mu \\
&= \frac{1}{2i\pi} \int_{a-i\infty}^{a+i\infty} (\zeta(\nu))^{-1} \mathfrak{E}_{-\nu} \, d\nu , \qquad a > 1 .
\end{aligned} \tag{12.62}$$

A change of contour finishes the proof. $\square$

Remark 12.1. It is easy to generalize the second part of Theorem 12.5 to the case when the function $A(r)$ is replaced, with the help of an arbitrary Dirichlet character χ, by the function

$$A^\chi(r) = \prod_{\substack{p \text{ prime} \\ p | r}} (1 - p \chi(p)) : \tag{12.63}$$

then, one sets

$$\mathfrak{T}_\infty^\chi(x, \xi) = \sum_{|j|+|k| \neq 0} A^\chi((j, k)) \delta(x - j) \delta(\xi - k) . \tag{12.64}$$

Since, with the same proof as in (12.51),

$$\sum_{r \geq 1} A^\chi(r) r^{-s} = \frac{\zeta(s)}{L(s - 1, \chi)} , \tag{12.65}$$

(12.46) generalizes as

$$\mathfrak{T}_\infty^\chi = \frac{1}{2\pi} \int_{-\infty}^{\infty} (L(-i\lambda, \chi))^{-1} \, \mathfrak{E}_{i\lambda} \, d\lambda + \sum_{L^*(\beta, \chi)=0} \mathrm{Res}_{\mu=\beta} \left(\frac{\mathfrak{E}_{-\mu}}{L(\mu, \chi)} \right), \qquad (12.66)$$

where, again, the equation $L^*(\beta, \chi) = 0$ characterizes the non-trivial zeros of the Dirichlet L-function with character χ.

Remark 12.2. Looking at the second equation (12.62), or at (12.46), it is clear that any piece of information regarding the structure of the distribution $\mathfrak{T}_\infty$ may be of interest. Indeed, note that the line of integration on the right-hand side of (12.62), or in the first term on the right-hand side of (12.46), can always be moved to a parallel line if one agrees to take into consideration the required residues at zeros of zeta. Then, the Riemann hypothesis just means that it is possible to write a decomposition of this distribution as a generalized integral (principal values, in Cauchy's sense, are needed because of the critical zeros of zeta) supported by the line $\mathrm{Re}\ \nu = \frac{1}{2}$. In view of (11.35), this phrasing can be easily transformed into a set of estimates about the automorphic functions obtained when testing $\mathfrak{T}_\infty$ against functions such as

$$W(u_z, u_z)(x, \xi) = 2 \exp \left(-2\pi \frac{|x - z\,\xi|^2}{\mathrm{Im}\ z} \right). \qquad (12.67)$$

Unsurprisingly, these estimates are out of reach, one of the reasons being the essential lack of uniformity with respect to N of the estimate in Lemma 12.2.

Let us consider the distribution $\mathfrak{T}_N$, of which $\mathfrak{T}_\infty$ is a limit, instead. From Theorem 12.1 and (12.42), one has

$$W_N^\chi = N^{i\pi\mathcal{E}} \prod_{p \in S} (1 - p^{-2i\pi\mathcal{E}}) \, [\mathfrak{D}_0 - \delta]. \qquad (12.68)$$

Using (11.5) and (12.43), one obtains

$$\mathfrak{T}_N = \prod_{p \in S} (1 - p^{-2i\pi\mathcal{E}}) \left[1 + \frac{1}{2\pi} \int_{-\infty}^{\infty} \mathfrak{E}_{i\lambda} \, d\lambda \right]$$

$$= \prod_{p \in S} (1 - p^{-1}) + \frac{1}{2\pi} \int_{-\infty}^{\infty} (\zeta_N(-i\lambda))^{-1} \, \mathfrak{E}_{i\lambda} \, d\lambda : \qquad (12.69)$$

using the information regarding the complex continuation of the distribution-valued function $\nu \mapsto \mathfrak{E}_\nu$ given between (11.1) and (11.2), this can be written as

$$\mathfrak{T}_N = \frac{1}{2\pi} \int_{-\infty}^{\infty} (\zeta_N(-i\lambda))^{-1} \, \mathfrak{E}_{i\lambda} \, d\lambda - \mathrm{Res}_{\mu=1} \left(\frac{\mathfrak{E}_{-\mu}}{\zeta_N(\mu)} \right), \qquad (12.70)$$

an expression with a strong formal similarity with (12.46).

When moving from $\mathfrak{T}_N$ to $\mathfrak{T}_\infty$ the following happens: first, the pole at $\mu = 1$ disappears because, contrary to ζ_N, ζ has a pole there; next, on the line $\text{Re } \mu = 0$, the function $\zeta_N(\mu)$ has no limit as $S \to P$, even though, in some formal sense, it looks reasonable that $\zeta(\mu)$ should appear; finally, a series of residues, at the non-trivial zeros of ζ, makes its appearance.

If all that precedes were based solely on the consideration of $\mathfrak{T}_\infty$ and $\mathfrak{T}_N$, defined as distributions in the plane, we would certainly regard it as insignificant. However, W_N, if not $\mathfrak{T}_N$, has another, quite different, interpretation as a sum of Wigner functions: we consider it as being potentially non-trivial since it depends on the metaplectic representation and on the arithmetically meaningful consideration of the set of squares in $(\mathbb{Z}/N\mathbb{Z})^\times$.

Remaining in the current analytic environment, the effect of the operator $N^{i\pi\mathcal{E}}$ which defines $\mathfrak{T}_N$ in terms of $W_N^\times$, while obvious so far as decompositions into homogeneous components are concerned, destroys the interpretation in terms of Wigner functions. The reason for this is that transformations $(x, \xi) \mapsto (\epsilon^{-1}x, \epsilon^{-1}\xi)$ do not lie in the symplectic group, and do not correspond to operations on functions on the line. Instead, their interpretation requires a "change of Planck's constant" (just as in semi-classical analysis, though ϵ, which stands for the usual overused letter h, has in our case little to do with the physicists' Planck's constant). Just so as to answer a possible question from the reader, let us mention that one defines the ϵ-dependent Weyl calculus by the equation

$$(\text{Op}_\epsilon(\mathfrak{S})\,u)(x) = \epsilon^{-1} \int_{\mathbb{R}^2} \mathfrak{S}(\frac{x+y}{2}, \eta)\, e^{\frac{2i\pi}{\epsilon}(x-y)\eta}\, u(y)\, dy\, d\eta\,, \tag{12.71}$$

in other words,

$$\text{Op}_\epsilon(\mathfrak{S}) = \text{Op}\left((y, \eta) \mapsto \mathfrak{S}(y, \epsilon\eta)\right). \tag{12.72}$$

Combine this with the covariance property of the Weyl calculus under rescaling transformations: with $(T_a\,u)(x) = a^{-\frac{1}{2}}\,u(a^{-1}x)$, one has $T_a\,\text{Op}(\mathfrak{S})\,T_a^{-1} = \text{Op}\left((y,\eta) \mapsto \mathfrak{S}(a^{-1}y, a\eta)\right)$. The result is that, given $\mathfrak{S} \in \mathcal{S}'(\mathbb{R}^2)$, the distribution $\epsilon^{-i\pi\mathcal{E}}\,\mathfrak{S}$ is $\epsilon^{-\frac{1}{2}}$ times the Op_ϵ-symbol of the operator $T_{\sqrt{\epsilon}}^{-1}\,\text{Op}(\mathfrak{S})\,T_{\sqrt{\epsilon}}$. Of course, you might wish to consider ϵ as being "small", which is not the case of N, but since W_N is invariant under the transformation $(-1)^{\#S}\,\mathcal{G}$ (cf. what follows (2.8)), Theorem 12.5 is equivalent to the version obtained after one has applied $\mathcal{G}$, an operation which changes $N^{-i\pi\mathcal{E}}$ to $N^{i\pi\mathcal{E}}$.

The remark just made about changing Planck's constant was only made for clarity: what it really implies is that applying the operator $N^{-i\pi\mathcal{E}}$ has destroyed the most interesting feature of the distribution W_N. This is not to mean that quantization theory is necessarily irrelevant towards the study of the zeta function, and we hope to come back to this point of view in the future.

13 Spaces of combs and Dirichlet series

In this section, we prepare the ground for future investigations, starting from a new interpretation of the full set of coefficients $\Gamma_{jk}(\rho_1, \rho_2)$: up to now, we have been concentrating on the sum of the diagonal ones (with respect to the ρ variables) only. This will take us naturally to the consideration of *combs*, introduced in [30, Section 16] as special automorphic distributions, modelled after the Dirac comb. They can be identified with the Dirichlet series made from their coefficients. It is possible, as will be shown towards the end of the book, to develop a kind of analysis specifically adapted to functions which arise as Dirichlet series. We certainly plan to explore this idea further in the future: the first few steps in this direction already reveal some puzzling facts.

With the usual assumption that N is 4 times a squarefree odd integer, recall (4.20) that

$$\Gamma_{jk}(\rho_1, \rho_2) = \frac{1}{2} \sum_{\substack{\mu_1, \mu_2 \in \Lambda \\ \langle \rho, \mu \rangle \equiv 2j \bmod N}} \chi(\mu_1 \mu_2) \, \exp\left(2i\pi k \, \frac{\langle J\rho, \mu \rangle}{N} \right). \qquad (13.1)$$

The definition of these coefficients makes sense whenever $\rho_1, \rho_2 \in (\mathbb{Z}/N\mathbb{Z})^\times$: there is no need to limit their definition to the case of a set of representatives mod Λ. To help intuition – and typography as well – let us use, in the finite-dimensional case, notation more usual in the case of function spaces. Denote as $\ell^2((\mathbb{Z}/N\mathbb{Z})^\times)$ the linear space of complex-valued functions on $(\mathbb{Z}/N\mathbb{Z})^\times$: the "integral kernel" of a linear endomorphism of this space (we shall drop the quotation marks from now on) is none other than the corresponding matrix $L = (L_{\rho_1, \rho_2})_{\rho_1, \rho_2 \in (\mathbb{Z}/N\mathbb{Z})^\times}$, an element of $\ell^2((\mathbb{Z}/N\mathbb{Z})^\times) \otimes \ell^2((\mathbb{Z}/N\mathbb{Z})^\times)$.

Note that this is true only if we use on this matrix space the measure which gives the mass 1 to each entry: such a measure is not self-dual, which does not matter in this finite-dimensional setting. Linear endomorphisms of the space $\ell^2((\mathbb{Z}/N\mathbb{Z})^\times)$ can also be defined by means of "symbols", in a way which we now make precise. These are functions on the set $\mathbb{Z}/N'\mathbb{Z} \times \mathbb{Z}/N'\mathbb{Z}$, with $N' = N/2$, in other words matrices too, of a different species. We denote the space of such functions as $\ell^2(\mathbb{Z}/N'\mathbb{Z}) \otimes \ell^2(\mathbb{Z}/N'\mathbb{Z})$, and choose on it the measure for which, again, each entry has mass 1.

Recall from (2.2) that, on the real line, the operator with Weyl symbol h is the operator with integral kernel

$$(x, y) \mapsto \int_{-\infty}^{\infty} h\left(\frac{x+y}{2}, \eta \right) e^{2i\pi(x-y)\eta} \, d\eta. \qquad (13.2)$$

Defining $\theta = \exp \frac{2i\pi}{N}$ as in (3.2), we are led to the following finite-dimensional analogue.

Definition 13.1. Given $h \in \ell^2(\mathbb{Z}/N'\mathbb{Z}) \otimes \ell^2(\mathbb{Z}/N'\mathbb{Z})$, the operator $\mathrm{Op}^{(N)}(h)$ with Weyl symbol h is the linear endomorphism of the space $\ell^2(\mathbb{Z}/N\mathbb{Z})$ the integral kernel (*i.e.*, the matrix) of which is the function

$$(\rho_1, \rho_2) \mapsto \frac{1}{2} \sum_{k \in \mathbb{Z}/N'\mathbb{Z}} h\left(\frac{\rho_1 + \rho_2}{2}, k\right) \theta^{(-\rho_1 + \rho_2)k}. \tag{13.3}$$

Restricting the function of (ρ_1, ρ_2) so defined to $(\mathbb{Z}/N\mathbb{Z})^\times \times (\mathbb{Z}/N\mathbb{Z})^\times$, one obtains the integral kernel of a linear endomorphism of the space $\ell^2\left((\mathbb{Z}/N\mathbb{Z})^\times\right)$, which we shall denote as $\mathrm{Op}_\times^{(N)}(h)$.

Immediately note that, since $4|N$, ρ_1 and ρ_2 are odd, and their sum is a well-defined even integer mod N, so that their half-sum is indeed well defined mod N'; on the other hand, since $\rho_1 - \rho_2$ is even too, the power of θ is indeed well defined as soon as one knows k mod N'. This gives (13.3) a meaning. The operator the symbol of which is the constant function 1 is $\frac{N}{4}$ times the identity: we do not feel concerned, here, with normalizations, which could always be fixed at the end if so desired.

Definition 13.2. For every $\mu \in \Lambda$, denote as $((\mu))$ the automorphism of $\ell^2\left((\mathbb{Z}/N\mathbb{Z})^\times\right)$, corresponding to a permutation of $(\mathbb{Z}/N\mathbb{Z})^\times$, consisting in substituting for a function u on $(\mathbb{Z}/N\mathbb{Z})^\times$ the function $\rho \mapsto u(\mu \rho)$. Given a symbol $h \in \ell^2(\mathbb{Z}/N'\mathbb{Z}) \otimes \ell^2(\mathbb{Z}/N'\mathbb{Z})$, consider the operator

$$B = 4^{-\#S} \sum_{\mu_1, \mu_2 \in \Lambda} \chi(\mu_1 \mu_2) \, ((\mu_1)) \, \mathrm{Op}_\times^{(N)}(h) \, ((\mu_2^{-1})), \tag{13.4}$$

a linear endomorphism of $\ell^2\left((\mathbb{Z}/N\mathbb{Z})^\times\right)$. The integral kernel of B will be denoted as $\mathbf{\Gamma}^* h$, thus defining a linear map $\mathbf{\Gamma}^*$ from $\ell^2(\mathbb{Z}/N'\mathbb{Z}) \otimes \ell^2(\mathbb{Z}/N'\mathbb{Z})$ to $\ell^2\left((\mathbb{Z}/N\mathbb{Z})^\times\right) \otimes \ell^2\left((\mathbb{Z}/N\mathbb{Z})^\times\right)$: recall that $\mu_2^{-1} = \mu_2$ since $\mu_2 \in \Lambda$.

The map $\mathbf{\Gamma}^*$ is exactly that which makes the interpretation of the coefficients $\Gamma_{jk}(\rho_1, \rho_2)$ we have in mind possible.

Theorem 13.3. *Given* $h \in \ell^2(\mathbb{Z}/N'\mathbb{Z}) \otimes \ell^2(\mathbb{Z}/N'\mathbb{Z})$, *the matrix* $L = \mathbf{\Gamma}^* h$ *is characterized by the equation*

$$L(\rho_1, \rho_2) = 4^{-\#S} \sum_{j, \, k \bmod N'} \overline{\Gamma}_{jk}(\rho_1, \rho_2) \, h(j, k). \tag{13.5}$$

Proof. Applying (13.3), letting $((\mu_1))$ and $((\mu_2^{-1}))$ act on both sides and summing, one obtains that the sought-after integral kernel is given as

$$(\rho_1, \rho_2) \mapsto 4^{-\#S} \times \frac{1}{2} \sum_{\mu_1, \, \mu_2 \in \Lambda} \chi(\mu_1 \mu_2) \sum_{k \in \mathbb{Z}/N'\mathbb{Z}} h\left(\frac{\langle \rho, \mu \rangle}{2}, k\right) \theta^{-\langle J\rho, \mu \rangle \, k} : \tag{13.6}$$

recall (4.9) for the definition of $\langle J \cdot, \cdot \rangle$ if needed. On the other hand,

$$4^{-\#S} \sum_{j,\, k \bmod N'} \overline{\Gamma}_{jk}(\rho_1, \rho_2)\, h(j,\, k)$$

$$= 4^{-\#S} \times \frac{1}{2} \sum_{j \bmod N'} \sum_{\substack{\mu_1,\, \mu_2 \in \Lambda \\ \langle \rho,\, \mu \rangle \equiv 2j \bmod N}} \chi(\mu_1 \mu_2) \sum_{k \bmod N'} \theta^{-k \langle J\rho,\, \mu \rangle}\, h(j,\, k) : \quad (13.7)$$

performing the summation with respect to j, we get back to the preceding expression. $\qquad\square$

It is easy to see that the image of the space $\ell^2(\mathbb{Z}/N'\mathbb{Z}) \otimes \ell^2(\mathbb{Z}/N'\mathbb{Z})$ under Γ^* is exactly the set of matrices $L \in \ell^2((\mathbb{Z}/N\mathbb{Z})^\times) \otimes \ell^2((\mathbb{Z}/N\mathbb{Z})^\times)$ satisfying the property that

$$L(\mu\rho_1, \rho_2) = L(\rho_1, \mu\rho_2) = \chi(\mu)\, L(\rho_1, \rho_2) \qquad \text{for every } \mu \in \Lambda. \quad (13.8)$$

Actually, if such is the case, one has $\Gamma^* \Gamma L = \frac{N}{4}\, L$, as shown by a short computation. Of course, Γ^* is far from being one-to-one since the image of this map has dimension $\prod_{3 \le p \in S} \frac{(p-1)^2}{4}$, a number smaller than p^2. After a rather lengthy calculation involving the decomposition, for every prime p, of the quasiregular action of $SL(2, \mathbb{Z}/p\mathbb{Z})$ on $\ell^2(\mathbb{Z}/p\mathbb{Z}) \otimes \ell^2(\mathbb{Z}/p\mathbb{Z})$ into irreducibles, one can see that $\Gamma \Gamma^*$ is $\frac{N}{4}$ times an orthogonal projection.

For a given N, the full set of coefficients $\Gamma_{jk}(\rho_1, \rho_2)$ involved in the calculations of Sections 4 and 12 is thus already brought to light by the comparison between two very natural ways to describe linear operators on $\ell^2((\mathbb{Z}/N\mathbb{Z})^\times)$. In order to consider all values of N simultaneously, it is necessary to imbed the two spaces of matrices $\ell^2(\mathbb{Z}/N'\mathbb{Z}) \otimes \ell^2(\mathbb{Z}/N'\mathbb{Z})$ (resp. $\ell^2((\mathbb{Z}/N\mathbb{Z})^\times) \otimes \ell^2((\mathbb{Z}/N\mathbb{Z})^\times)$) into much larger spaces. Spaces of complex-valued functions on two-dimensional spaces of adeles (resp. ideles) would seem to be quite appropriate since, in order to approach the zeta function by the partial products of its Eulerian expansion, we just have to let the set of prime divisors of N go to P, the set of all primes. However, so far at least as the first of these two spaces is concerned, calculations relative to Wigner functions took us to another solution, making for every pair (ρ_1, ρ_2) the distribution $\mathfrak{S}_{\rho_1, \rho_2}$ in (4.25) out of the set of coefficients $(\Gamma_{jk}(\rho_1, \rho_2))_{j,k \bmod N'}$.

Define *combs* [30, Section 16] (in [32], the theory is applied to generalizations of Voronoï's formula [34]) as being measures on $\mathbb{R}^2$ supported in $\mathbb{Z}^2 \backslash \{0\}$, invariant under the linear action of $SL(2, \mathbb{Z})$. In other words, these are distributions $\mathfrak{S}$ such that

$$\mathfrak{S}(x, \xi) = \sum_{|j| + |k| \ne 0} a((j,\, k))\, \delta(x - j)\, \delta(\xi - k), \quad (13.9)$$

where $a(r)$ is some function of $r = 1, 2, \ldots$ and, of course, $(j,\, k) = \text{g.c.d.}(j,\, k)$: we also assume that $\mathfrak{S}$ is a tempered distribution, *i.e*, that $a(r)$ is bounded by some power of $1 + r$. We shall sometimes denote as $\mathfrak{a}$ the sequence $(a(r))_{r \ge 1}$.

Provided one extends the notion a little bit, which will be done in a moment, what one obtains is equivalent to that of an appropriate function on the two-dimensional adelic space. As long as one does not use any sophisticated concept from algebraic number theory, choosing between a description in terms of automorphic distribution theory or an adelic description seems to be a matter of taste. As a first exercise, let us show how the Möbius inversion formula makes it possible to circulate between various combs, at the same time putting forward the role of the comb $\mathfrak{T}_\infty$ as a generating object. Needless to say, one could use just as well the comb $\mathfrak{T}_\infty^\chi$ defined in Remark 12.1 in connection with a Dirichlet character χ.

Proposition 13.4. *Given a comb $\mathfrak{S}$ as defined in (13.9), define the functions g and f on the set of positive integers by the equations*

$$\frac{g(r)}{r} = \sum_{1 \le \delta \mid r} \frac{a(\delta)}{\delta},$$

$$r\,f(r) = \sum_{1 \le d \mid r} \text{Möb}(d)\,g\!\left(\frac{r}{d}\right). \tag{13.10}$$

Then, introducing the Dirichlet series $F(s) = \sum_{r \ge 1} \frac{f(r)}{r^s}$, one has

$$\mathfrak{S} = F(2i\pi\mathcal{E})\,\mathfrak{T}_\infty. \tag{13.11}$$

Proof. Before we give the easy proof, let us emphasize that, when acting on combs, Dirichlet series with the Euler operator $2i\pi\mathcal{E}$ as an argument are especially easy to deal with. Indeed, writing

$$F(2i\pi\mathcal{E}) = \sum_{d \ge 1} f(d)\,d^{-2i\pi\mathcal{E}}, \tag{13.12}$$

and using (4.23) to the effect that

$$d^{-2i\pi\mathcal{E}}\,(\delta(x-j)\,\delta(\xi-k)) = d\,\delta(x-dj)\,\delta(\xi-dk), \tag{13.13}$$

one sees that the family of supports of transforms, under the operators $d^{-2i\pi\mathcal{E}}$, of any comb, is locally finite. In particular, starting from the equation

$$\mathfrak{T}_\infty(x,\xi) = \sum_{|j|+|k|\neq 0} A((j,k))\,\delta(x-j)\,\delta(\xi-k), \tag{13.14}$$

with $A(r)$ as made explicit in (12.48), one sees that (13.11) will hold provided that

$$\mathfrak{S}(x,\xi) = \sum_{d \ge 1} d\,f(d) \sum_{|j|+|k|\neq 0} A((j,k))\,\delta(x-dj)\,\delta(\xi-dk) \tag{13.15}$$

or, comparing to (13.9),

$$a((j,\, k)) = \sum_{1 \le d | (j,k)} d\, f(d)\, A\left(\left(\frac{j}{d},\, \frac{k}{d}\right)\right),$$ (13.16)

which is the same as the equation

$$a(r) = \sum_{1 \le d | r} d\, f(d)\, A\left(\frac{r}{d}\right).$$ (13.17)

In view of (12.50), this is equivalent to

$$a(r) = \sum_{1 \le d | r} d\, f(d) \sum_{1 \le \delta | \frac{r}{d}} \delta\, \mathrm{M\ddot{o}b}(\delta),$$ (13.18)

or

$$a(r) = \sum_{1 \le \delta | r} g\left(\frac{r}{\delta}\right) \delta\, \mathrm{M\ddot{o}b}(\delta)$$ (13.19)

if g is defined by the equation

$$g(r) = \sum_{1 \le d | r} d\, f(d).$$ (13.20)

Now, the Möbius inversion formula, applied to (13.19), yields the first equation (13.10); when applied to (13.20), it yields the second equation (13.10). $\square$

If one starts with a small space $\mathfrak{C}_0$ of combs as defined by (13.11), for instance those obtained when the Dirichlet series $F(s)$ extends as a holomorphic function in a neighbourhood of the half-plane $\mathrm{Re}\ s \ge \frac{1}{2}$, satisfying the condition, usual in such matters, of being polynomially bounded in vertical strips, one finds, starting from (12.62), the equation

$$\mathfrak{S} = \frac{1}{2i\pi} \int_{\frac{1}{2}-i\infty}^{\frac{1}{2}+i\infty} \frac{F(\nu)}{\zeta(\nu)}\, \mathfrak{E}_{-\nu}\, d\nu + \sum_{\substack{\zeta^*(\beta)=0 \\ \mathrm{Re}\ \beta > \frac{1}{2}}} \mathrm{Res}_{\mu=\beta} \left(\frac{F(\mu)}{\zeta(\mu)}\, \mathfrak{E}_{-\mu}\right) :$$ (13.21)

it is understood that the integral is the arithmetic means of the two ones obtained by changing slightly the contour of integration so as to bypass all critical zeros of zeta (but no other, if any such should exist), leaving those either on the left or on the right of the new contour.

Then, of course, one may think of trying to attack R.H. by extending $\mathfrak{C}_0$ as a Hilbert space on which the operator $i(\frac{1}{2} - 2i\pi\mathcal{E})$ will have a self-adjoint extension, for which the distributions $\mathfrak{E}_{-\beta}$ with $\zeta^*(\beta) = 0$ would be eigenfunctions: let us hasten to say that the considerations that follow do not succeed in this respect. The pre-Hilbert spaces of combs to be discussed in what follows, even when completed, leave out all individual distributions $\mathfrak{E}_{-\beta}$. Instead of looking at the operator $i(\frac{1}{2} - 2i\pi\mathcal{E})$, one may look at the family of operators $p^{\frac{1}{2} - 2i\pi\mathcal{E}}$, p prime, which, as already seen, are easy to describe when acting on combs, and ask that they should be unitary. Combs, however, may be too restricted a notion: the extension from combs to fractional combs, to be introduced presently, has considerable similarity to that which yields the adele ring, starting from the subring consisting of adeles which are integral at every non-archimedean place.

Definition 13.5. A fractional comb $\mathfrak{S}$ is any measure of the kind

$$\mathfrak{S}(x, \xi) = \sum_{\substack{\alpha, \beta \in \mathbb{Q} \\ |\alpha| + |\beta| \neq 0}} C(\alpha, \beta)\, \delta(x - \alpha)\, \delta(\xi - \beta), \qquad (13.22)$$

with the following property: there exists $M = 1, 2, \ldots$ such that $C(\alpha, \beta) = 0$ unless $M\alpha$ and $M\beta$ are integers; moreover, for some such choice of M, $C(\alpha, \beta)$ depends only on $(M\alpha, M\beta) = \text{g.c.d.}\,(M\alpha, M\beta)$.

Note that if M satisfies the condition above, in which case we shall say that it is adapted to $\mathfrak{S}$, so does kM for every integer $k \geq 1$ since $(M\alpha, M\beta) = \frac{(kM\alpha, kM\beta)}{k}$: this shows in particular that the set of fractional combs is a linear space. Moreover, if one defines the function $a^{(M)}$ on the set of positive integers so that

$$C(\alpha, \beta) = a^{(M)}((M\alpha, M\beta)) \qquad (13.23)$$

whenever $M\alpha$ and $M\beta$ are integers, one has for every $r \geq 1$ the identity

$$a^{(kM)}(r) = \begin{cases} a^{(M)}(\frac{r}{k}) & \text{if } k | r, \\ 0 & \text{otherwise}. \end{cases} \qquad (13.24)$$

On the other hand, the fractional comb

$$\mathfrak{S}^{(\tau)} = M^{\tau - 2i\pi\mathcal{E}}\, \mathfrak{S} \qquad (13.25)$$

is actually a comb (it also depends on M unless we choose M "minimal"). Its coefficients $a((j, k))$, defined just as in (13.9), are given by the equation $a(r) = M^{\tau+1} a^{(M)}(r)$. This makes it possible to extend all computations relative to combs to fractional combs in a trivial way: for simplicity, we shall deal with combs only in all that follows.

Let us remark that the class of fractional combs is not invariant under the Fourier transformation: it becomes so if further restricted by the constraint that, for some integer $M' \geq 1$, $a^{(M)}(r)$ should depend only on the class of r mod M'. However, such a definition would exclude the distribution $\mathfrak{T}_\infty$ (but not $\mathfrak{T}_N$ for finite N).

Defining a comb is fully equivalent to defining its sequence $\mathfrak{a} = (a(m))_{m \geq 1}$ of coefficients, which can be any polynomially bounded sequence of complex numbers.

Then, one can characterize it by the associated Dirichlet series

$$L(s, \mathfrak{a}) = \sum_{m \geq 1} \frac{a(m)}{m^s}, \tag{13.26}$$

convergent for Re s large enough. We shall also use the functions

$$L^*(s, \mathfrak{a}) = \pi^{-\frac{s}{2}} \Gamma(\frac{s}{2}) L(s, \mathfrak{a}), \qquad L^{**}(s, \mathfrak{a}) = (2\pi)^{-\frac{s}{2}} \Gamma(s) L(s, \mathfrak{a}). \tag{13.27}$$

It is our feeling that one should develop an analysis adapted precisely to the study of Dirichlet series. It should, as a start, include such technical tools as the definition of Hilbert spaces of Dirichlet series, a pseudodifferential analysis, a connection between the additive and multiplicative theories.... Moving back and forth between the point of view of combs and that of Dirichlet series cannot fail to help in this direction. The end of the book is devoted to what could be the rudiments of such a theory. Let us start with the following object:

Definition 13.6. Given two polynomially bounded sequences $\mathfrak{a} = (a(m))_{m \geq 1}$ and $\mathfrak{b} = (b(m))_{m \geq 1}$ we define, when Re s is large and $y > 0$, the expression

$$W[\mathfrak{a}, \mathfrak{b}](s; iy) = \frac{1}{4\pi} \int_{-\infty}^{\infty} L^*(s - i\lambda, \mathfrak{a}) L^*(s + i\lambda, \mathfrak{b}) y^{i\lambda} \, d\lambda. \tag{13.28}$$

The reader will note the strong analogy with the Wigner function as defined in (2.3), as applied to the pair of Dirichlet series associated with the two sequences: only, instead of y, it is the logarithm of y that would be considered as the natural variable there. There are a few other minor differences too: it is in the present context less disconcerting to use a bilinear, rather than a sesquilinear, form. Last, our choice of emphasizing iy rather than y is not fortuitous: indeed, the function $iy \mapsto W[\mathfrak{a}, \mathfrak{b}](s; iy)$ has a natural extension as a function in the upper half-plane.

Theorem 13.7. *Recall* [16, p. 166] *that the Legendre function on the cut* $\,] - 1, 1[$ *is that defined in terms of the hypergeometric function by the equation*

$$\left(\frac{1 - c}{1 + c}\right)^{\frac{1}{4} - \frac{s}{2}} P_{\nu}^{\frac{1}{2} - s}(c) = \left(\Gamma \left(\frac{1}{2} + s \right) \right)^{-1} {}_2F_1 \left(-\nu, 1 + \nu; \frac{1}{2} + s; \frac{1 - c}{2} \right). \tag{13.29}$$

Under the assumptions of the definition that precedes, and for Re s *large enough, one has for every point* $z \in \Pi$ *the identity*

$$\pi^{-s} \Gamma(s) \sum_{m, n \geq 1} a(m) b(n) \left(\frac{|mz - n|^2}{\text{Im } z} \right)^{-s} \tag{13.30}$$

$$= (2\pi)^{-\frac{1}{2}} (\text{Im } z)^{\frac{1}{2}} \int_{-\infty}^{\infty} L^{**}(s - i\lambda, \mathfrak{a}) L^{**}(s + i\lambda, \mathfrak{b}) |z|^{-\frac{1}{2} + i\lambda} P_{-\frac{1}{2} + i\lambda}^{\frac{1}{2} - s} \left(-\frac{\text{Re } z}{|z|} \right) d\lambda.$$

The function $F(z)$ so defined is the extension of the function $iy \mapsto W[\mathfrak{a}, \mathfrak{b}](s; iy)$ as an eigenfunction on Π of the hyperbolic Laplacian, to wit the solution of the Cauchy problem

$$\Delta F = s(1-s) F,$$
$$F(iy) = W[\mathfrak{a}, \mathfrak{b}](s; iy),$$
$$\left.\frac{\partial}{\partial x}\right|_{x=0} F(x+iy) = -\frac{1}{\pi} \int_{-\infty}^{\infty} \frac{\Gamma(\frac{1+s-i\lambda}{2})}{\Gamma(\frac{s-i\lambda}{2})} L^*(s-i\lambda, \mathfrak{a}) L^*(s+i\lambda, \mathfrak{b}) y^{-1+i\lambda} d\lambda.$$

$$(13.31)$$

Before giving the proof, we need a lemma.

Lemma 13.8. *Assuming* $-1 < c < 1$ *and* Re $s > 0$, *one has*

$$(\cosh \pi\xi + c)^{-s} = \frac{(2\pi)^{-\frac{1}{2}}}{\Gamma(s)} (1-c^2)^{\frac{1}{4}-\frac{s}{2}} \int_{-\infty}^{\infty} e^{i\pi\xi\lambda} \Gamma(s-i\lambda) \Gamma(s+i\lambda) P^{\frac{1}{2}-s}_{-\frac{1}{2}+i\lambda}(c) d\lambda.$$

$$(13.32)$$

When $s = \sigma$ *is real* $> \frac{1}{2}$, $P^{\frac{1}{2}-\sigma}_{-\frac{1}{2}+i\lambda}(c)$ *is positive.*

Proof. From [16, p. 188], one has, if $-1 < c < 1$ and Re $s > |\frac{1}{2} + $ Re $\nu|$, the equation

$$\Gamma(s) \int_0^{\infty} (\cosh \pi\xi + c)^{-s} \cosh\left(\left(\nu + \frac{1}{2}\right)\xi\right) d\xi$$
$$= \left(\frac{\pi}{2}\right)^{\frac{1}{2}} \Gamma\left(s - \frac{1}{2} - \nu\right) \Gamma\left(s + \frac{1}{2} + \nu\right) (1-c^2)^{\frac{1}{4}-\frac{s}{2}} P^{\frac{1}{2}-s}_{\nu}(c), \quad (13.33)$$

hence, for Re $s > 0$,

$$\int_0^{\infty} (\cosh \pi\xi + c)^{-s} \cos \xi\eta \, d\xi = \left(\frac{\pi}{2}\right)^{\frac{1}{2}} \frac{\Gamma(s-i\eta)\Gamma(s+i\eta)}{\Gamma(s)} (1-c^2)^{\frac{1}{4}-\frac{s}{2}} P^{\frac{1}{2}-s}_{-\frac{1}{2}+i\eta}(c),$$

$$(13.34)$$

which implies the first part of the lemma, applying the Fourier inversion formula.

The second part is a consequence of the equation [16, p. 190]

$$\Gamma\left(s - \frac{1}{2}\right) P^{\frac{1}{2}-\sigma}_{-\frac{1}{2}+it}(c) = (1-c^2)^{\frac{1}{4}-\frac{\sigma}{2}} \int_c^1 (x-c)^{\sigma-\frac{3}{2}} P_{-\frac{1}{2}+it}(x) \, dx, \quad (13.35)$$

since [16, p. 167]

$$P_{-\frac{1}{2}+it}(x) = {}_2F_1\left(\frac{1}{4} + \frac{it}{2}, \frac{1}{4} - \frac{it}{2}; 1; 1 - x^2\right)$$
$$= \sum_0^{\infty} \frac{(\frac{1}{4} + \frac{it}{2})_n (\frac{1}{4} - \frac{it}{2})_n}{(n!)^2} (1-x^2)^n, \quad (13.36)$$

where we have used the Pochammer symbols $(\alpha)_n = \alpha(\alpha+1)\cdots(\alpha+n-1)$. $\square$

Proof of Theorem 13.7. Writing

$$\frac{|mz - n|^2}{2|z|\,mn} = \frac{1}{2}\left(\frac{|z|m}{n} + \frac{n}{|z|m}\right) - \frac{\operatorname{Re} z}{|z|}, \tag{13.37}$$

one can apply (13.32) with $\frac{|z|m}{n} = e^{\pi\xi}$, $c = -\frac{\operatorname{Re} z}{|z|}$, obtaining

$$\left(\frac{|mz - n|^2}{2|z|\,mn}\right)^{-s} = \frac{(2\pi)^{-\frac{1}{2}}}{\Gamma(s)}\left(\frac{\operatorname{Im} z}{|z|}\right)^{\frac{1}{2}-s}$$

$$\int \left(\frac{|z|m}{n}\right)^{i\lambda} \Gamma(s - i\lambda)\,\Gamma(s + i\lambda)\,P_{-\frac{1}{2}+i\lambda}^{\frac{1}{2}-s}\left(-\frac{\operatorname{Re} z}{|z|}\right) d\lambda, \tag{13.38}$$

and equation (13.33) follows from (13.27). Computing the first two traces of the function F on the pure imaginary half-line is easy, in view of the equations [16, p. 171]

$$P_{-\frac{1}{2}+i\lambda}^{\frac{1}{2}-s}(0) = \frac{2^{\frac{1}{2}-s}\,\pi^{\frac{1}{2}}}{\Gamma(\frac{1+s+i\lambda}{2})\,\Gamma(\frac{1+s-i\lambda}{2})},$$

$$(P_{-\frac{1}{2}+i\lambda}^{\frac{1}{2}-s})'(0) = -\frac{2^{\frac{3}{2}-s}\,\pi^{\frac{1}{2}}}{\Gamma(\frac{1+s+i\lambda}{2})\,\Gamma(\frac{s-i\lambda}{2})} \tag{13.39}$$

and after some calculations involving the duplication formula for the Gamma function.   □

Considering the case when $\mathfrak{b} = \bar{\mathfrak{a}}$, one sees that Theorem 13.7 provides the construction of a two-parameter family of pre-Hilbert norms on suitable spaces of combs, or Dirichlet series. Indeed, if one takes for s any real number $\sigma > \frac{1}{2}$, and if one takes for z any point of Π on the circle $|z| = 1$, the complex measure $|z|^{-\frac{1}{2}+i\lambda}\,P_{-\frac{1}{2}+i\lambda}^{\frac{1}{2}-s}\left(-\frac{\operatorname{Re} z}{|z|}\right) d\lambda$ on the right-hand side of (13.33) becomes a positive measure, in view of the second part of Lemma 13.8.

Remark 13.1. In general, the left-hand side of (13.33) has of course no property of automorphy. Under the transformation (associated with a unimodular matrix) $z \mapsto \frac{1-z}{1+z}$, the imaginary half-line is sent to the half-circle with diameter $(-1, 1)$, but these two sets of values of z play very different roles here: the first one is associated to the notion of Wigner function (adapted to Dirichlet series), while the second one is that which provides positive-definite hermitian forms on spaces of combs, or Dirichlet series.

When $z = i$, the integral on the right-hand side of (13.33) simplifies as a consequence of the first equation (13.39). The identity which is the object of Theorem 13.7 becomes

$$\sum_{m,n\geq 1} \frac{\overline{a(m)}\,a(n)}{(m^2 + n^2)^{s+1}} = \frac{1}{4\pi}\int_{-\infty}^{\infty} L^*(s + i\lambda, \bar{\mathfrak{a}})\,L^*(s - i\lambda, \mathfrak{a})\,d\lambda. \tag{13.40}$$

In this case, there is a shorter proof of the identity, based on the equation (use the Fourier inversion formula, starting from [16, p. 407])

$$(\cosh \pi \xi)^{-s} = \frac{2^s}{4\pi \, \Gamma(s)} \int_{-\infty}^{\infty} e^{i\pi\xi\lambda} \, \Gamma\left(\frac{s+i\lambda)}{2}\right) \Gamma\left(\frac{s-it)}{2}\right) d\lambda, \qquad \mathrm{Re}\ s > 0,$$
$$(13.41)$$

which implies, if $m, n \geq 1$,

$$\int_{-\infty}^{\infty} \left(\frac{n}{m}\right)^{i\lambda} \Gamma\left(\frac{s+i\lambda)}{2}\right) \Gamma\left(\frac{s-i\lambda)}{2}\right) d\lambda = 4\pi\, \Gamma(s) \left(\frac{n}{m} + \frac{m}{n}\right)^{-s}, \qquad (13.42)$$

and leads to the required formula.

We now concentrate on this case, for simplicity only. Still, the parameter τ is left free.

Proposition 13.9. *Let $\tau > -1$ be given. Given a comb $\mathfrak{S}$ with coefficients $a(r)$, $r \geq 1$, as in (13.9), set, provided absolute convergence is ensured,*

$$(\mathfrak{S} \,\|\, \mathfrak{S})_{\tau+1} = \sum_{m,n \geq 1} \frac{\overline{a(m)}\, a(n)}{(m^2 + n^2)^{\tau+1}}. \qquad (13.43)$$

The hermitian form so defined is invariant under all operators $k^{\tau - 2i\pi\mathcal{E}}$, $k = 1, 2, \ldots$. It is positive-definite when restricted to the space of combs such that $\sum_{m \geq 1} \frac{|a(m)|}{m^{\tau+1}} < \infty$.

Proof. The invariance of the hermitian form under the given operators is a consequence of the equation

$$a'(r) = \begin{cases} k^{\tau+1}\, a(\frac{r}{k}) & \text{if } k|r, \\ 0 & \text{otherwise} \end{cases} \qquad (13.44)$$

that defines the set of coefficients $a'(r)$ of the comb $k^{\tau - 2i\pi\mathcal{E}}\, \mathfrak{S}$ in terms of those of the comb $\mathfrak{S}$. That the hermitian form is positive-definite is a consequence of (13.40).   □

Let us now interpret the hermitian form $(\mathfrak{S} \,\|\, \mathfrak{S})_{\tau+1}$ in terms of an appropriate spectral decomposition of $\mathfrak{S}$.

Proposition 13.10. *Assume that $-1 < \tau < 1$, and that the function $L(s, \mathfrak{a})$, defined for $\mathrm{Re}\ s$ large enough, extends as a holomorphic function in a neighbourhood of the half-plane $\mathrm{Re}\ s \geq \tau + 1$, bounded by some power of $1 + |\mathrm{Im}\ s|$ on any strip $\{s \in \mathbb{C} : \tau + 1 \leq \mathrm{Re}\ s \leq b < \infty\}$. Then, one has in a weak sense in $S'_{\mathrm{even}}(\mathbb{R}^2)$ the decomposition*

$$\mathfrak{S} = \frac{6}{\pi^2} L(2, \mathfrak{a}) + \frac{1}{2\pi} \int_{-\infty}^{\infty} \Phi_\tau(\lambda)\, \mathfrak{E}_{-\tau+i\lambda}\, d\lambda \qquad (13.45)$$

with

$$\Phi_\tau(\lambda) = \frac{L(\tau + 1 - i\lambda, \mathfrak{a})}{\zeta(\tau + 1 - i\lambda)}. \qquad (13.46)$$

In terms of the spectral density $\Phi_\tau(\lambda)$, *one has*

$$(\mathfrak{S} \parallel \mathfrak{S})_{\tau+1} = \frac{\pi^\tau}{4\,\Gamma(\tau+1)} \int_{-\infty}^\infty |L^*(\tau+1-i\lambda,\mathfrak{a})|^2 \, d\lambda$$

$$= \frac{\pi^\tau}{4\,\Gamma(\tau+1)} \int_{-\infty}^\infty |\Phi_\tau(\lambda)|^2 \, |\zeta^*(-\tau+i\lambda)|^2 \, d\lambda. \qquad (13.47)$$

Proof. In a way fully similar to the computations between (12.54) and (12.62), one obtains for large a the decomposition

$$\mathfrak{S} = \frac{1}{2i\pi} \int_{a-i\infty}^{a+i\infty} \frac{L(\nu+1,\mathfrak{a})}{\zeta(\nu+1)} \, \mathfrak{E}_{-\nu} \, d\nu. \qquad (13.48)$$

The decomposition (13.45) is obtained by a change of contour, using the fact that $\mathrm{Res}_{\nu=1}(\mathfrak{E}_{-\nu}) = 1$. The formula (13.47) follows from (13.46) and (13.27). $\square$

The scalar product under consideration has been built precisely so that the operator $i(\tau - 2i\pi\mathcal{E})$ should become a symmetric operator: it is therefore not surprising that the spectral decomposition should have a main integral term supported on the line that occurs in (13.45). In the case when $\tau = 0$, it is the operator $\mathcal{E}$ itself which has to be symmetric. In this case, there already exists [31, p. 30] a pre-Hilbert (incomplete) structure on the space of Γ-automorphic distributions with $\Gamma = SL(2,\mathbb{Z})$, with a norm denoted as $\| \cdot \|_\Gamma$.

The way this latter scalar product was constructed had to do, again, with the Weyl calculus. It is useful, in this context, to use in place of Op the modified version $\mathrm{Op}_{\sqrt2}$ defined by the equation

$$\mathrm{Op}_{\sqrt2}(\mathfrak{S}) = \mathrm{Op}\left((x,\xi) \mapsto \mathfrak{S}(2^{\frac12}x, 2^{\frac12}\xi)\right). \qquad (13.49)$$

With the help of the two families of coherent states (u_z) and (u_z^1) from Section 2, the norm was defined by the formula

$$\|\mathfrak{S}\|_\Gamma^2 = \| z \mapsto (u_z \,|\, \mathrm{Op}_{\sqrt2}(\mathfrak{S})\, u_z) \|_{L^2(\Gamma\backslash\Pi)}^2$$

$$+ \frac14 \left\| |\Delta - \tfrac14|^{-\frac12} \left(z \mapsto (u_z \,|\, \mathrm{Op}_{\sqrt2}(\mathfrak{S})\, u_z) \right) \right\|_{L^2(\Gamma\backslash\Pi)}^2. \qquad (13.50)$$

Here, $L^2(\Gamma\backslash\Pi)$ is the Hilbert space of automorphic functions in the upper half-plane Π square-integrable in the fundamental domain, and Δ is the standard self-adjoint realization of the hyperbolic Laplacian in $L^2(\Gamma\backslash\Pi)$: recall that the spectrum of Δ lies in $[\frac14, \infty[$, with the sole exception of the eigenvalue 0 (the corresponding eigenfunctions are the constants); the absolute value around $\Delta - \frac14$ takes care of this exceptional eigenvalue. Using [31, p. 30] together with (13.45), one obtains

$$\frac{\pi}{8} \|\mathfrak{S}\|_\Gamma^2 = \frac{3}{\pi^2} |L(2,\mathfrak{a})|^2 + (\mathfrak{S} \parallel \mathfrak{S})_0 : \qquad (13.51)$$

this formula connects the hermitian form obtained from Proposition 13.4 when $\tau = 0$ to the usual scalar product on automorphic functions in the upper half-plane. Contrary to the other Hilbert spaces considered here, or rather obtained under completion, the one considered in [31] also contained discrete (cusp-) eigendistributions, making the theory essentially equivalent – actually, slightly more precise – to that of the automorphic Laplacian in $L^2(\Gamma \backslash \Pi)$. We do not know whether anything similar can be done *in a useful way* when $\tau \neq 0$: of course, the discrete eigendistributions would no longer correspond to non-holomorphic cusp-forms, rather to Eisenstein distributions with non-trivial zeros of zeta on the appropriate line as parameters.

When $\mathfrak{S} = \mathfrak{T}_\infty$, one has $A(r) = \prod_{p|r}(1 - p)$ and, from (12.51), $L(s, A) = \frac{\zeta(s)}{\zeta(s-1)}$: in particular, the constant term $L(2, A)$ on the right-hand side of (13.45) is zero. From (13.47), the function

$$\tau \mapsto (\mathfrak{T}_\infty \| \mathfrak{T}_\infty)_{\tau+1} = \sum_{m,n \geq 1} \frac{A(m)\, A(n)}{(m^2 + n^2)^{\tau+1}} \tag{13.52}$$

extends as a holomorphic function of s in the half-plane

$$\mathrm{Re}\ s > \tau_0 = \sup\{\mathrm{Re}\ \beta: \zeta(\beta) = 0\} \in \left[\frac{1}{2}, 1\right],$$

but not in any bigger half-plane: it goes to $+\infty$ as $\tau \to \tau_0$. Just as another example, one sees that the series that corresponds to the choice of coefficients $a(r) = \mathrm{Möb}(r)$ extends as a holomorphic function of s in the half-plane $\mathrm{Re}\ s > \tau_0 - 1$, and goes to ∞ as $\tau \to \tau_0 - 1$. More generally, the analytic continuation, with respect to s, of series such as $\sum_{m,n \geq 1} \overline{a(m)}\, a(n)\, (m^2 + n^2)^{-s}$ associated to Dirichlet series $L(s, a)$ with a meromorphic extension to the complex plane, and with at most polynomial increase on vertical strips, has the following interesting feature: assuming that the poles of $L(s, a)$ are simple, the series under consideration extends an an analytic function in the complement of the closure Σ of the set of half-sums $\frac{1}{2}(\beta_1 + \overline{\beta}_2)$, where β_1 and β_2 are poles of the function $L(s, a)$. Indeed, if $s = \sigma + it \notin \Sigma$ but $s - i\lambda_0$ is a pole of $L(\cdot, a)$, then $s + i\lambda_0$ is not a pole of $L(\cdot, \overline{a})$, and in order to analyze the contribution to the integral of $L^*(s - i\lambda, a)\, L^*(s + i\lambda, \overline{a})$ (*cf.* (13.47)) of a small neighbourhood of λ_0, one can write this product as

$$L^*(s - i\lambda, a)\, L^*(s + i\lambda_0, \overline{a}) + (\lambda - \lambda_0)\, L^*(s - i\lambda, a)\, f(s, i\lambda) \tag{13.53}$$

with

$$L^*(s + i\lambda, \overline{a}) - L^*(s + i\lambda_0, \overline{a}) = (\lambda - \lambda_0)\, f(s, i\lambda) \tag{13.54}$$

and perform a slight change of contour of integration in λ in relation to the first term on the right-hand side of (13.53).

Index

Bibliography

[1] L. Auslander, L. Geshwind, F. Warner, *Weil multipliers*, J. Fourier Anal. Appl. **2**(2) (1995), 191–215.

[2] F.A. Berezin, *Quantization in Complex Symmetric Spaces*, Math. U.S.S.R. Izvestija **9**(2) (1975), 341–379.

[3] F.A. Berezin, *A connection between co- and contravariant symbols of operators on classical complex symmetric spaces*, Soviet Math. Dokl. **19** (1978), 786–789.

[4] D. Bump, *Automorphic Forms and Representations*, Cambridge Series in Adv. Math. **55**, Cambridge, 1996.

[5] P. Cartier, *An introduction to zeta functions*, in *From Number Theory to Physics*, (M. Waldschmidt, P. Moussa, J.M. Luck and C. Itzykson, eds.), Springer-Verlag, Berlin (1992).

[6] K. Chandrasekharan, *Arithmetical Functions*, Springer-Verlag, Berlin-Heidelberg-New York, 1970.

[7] K. Chandrasekharan, *Elliptic Functions*, Springer-Verlag, Berlin-Heidelberg-New York, 1985.

[8] J.M. Deshouillers, H. Iwaniec, *Kloosterman sums and Fourier coefficients of cusp forms*, Inv. Math. **70** (1982), 219–288.

[9] D. Goldfeld, P. Sarnak, *Sums of Kloosterman sums*, Inv. Math. **71**(2) (1983), 243–250.

[10] I.S. Gradstein, I.M. Ryshik, *Tables of Series, Products and Integrals*, vol. 2, Verlag Harri Deutsch, Thun-Frankfurt/M, 1981.

[11] S. Helgason, *Geometric Analysis on Symmetric Spaces*, Math. Surveys and Monographs **39**, A.M.S., Providence, 1994.

[12] H. Iwaniec, *Introduction to the spectral theory of automorphic forms*, Revista Matemática Iberoamericana, Madrid, 1995.

[13] H. Iwaniec, *Topics in Classical Automorphic Forms*, Graduate Studies in Math. **17**, A.M.S., Providence, 1997.

[14] H. Iwaniec, E. Kowalski, *Analytic Number Theory*, A.M.S. Colloquium Pub. **53**, Providence, 2004.

[15] P.D. Lax, R.S. Phillips, *Scattering Theory for Automorphic Functions*, Ann. Math. Studies **87**, Princeton Univ. Press, 1976.

[16] W. Magnus, F. Oberhettinger, R.P. Soni, *Formulas and theorems for the special functions of mathematical physics*, 3$^{\rm rd}$ edition, Springer-Verlag, Berlin, 1966.

[17] M.R. Murty, *Introduction to p-adic Aanalytic Number Theory*, Studies in Adv. Math. **27**, A.M.S., Providence, 2002.

[18] A. Ogg, *Modular Forms and Dirichlet Series*, Benjamin Inc., New York-Amsterdam, 1969.

[19] H. Rademacher, *Topics in Analytic Number Theory*, Springer-Verlag, Berlin, 1973.

[20] R.A. Rankin, *Modular Forms and Functions*, Cambridge Univ. Press, Cambridge-London-New York-Melbourne, 1977.

[21] A. Selberg, *On the Estimation of Fourier Coefficients of Modular Forms*, Proc. Symp. Pure Math. **8** (1963), 1–15.

[22] J.P. Serre, *Cours d'Arithmétique*, Presses Univ. de France, Paris, 1970.

[23] G. Shimura, *Modular Forms of half-integral weight*, Lecture Notes in Math. **320**, Springer-Verlag, Berlin-Heidelberg-New York, 1973.

[24] A. Terras, *Harmonic analysis on symmetric spaces and applications* 1, Springer-Verlag, New York-Berlin-Heidelberg-Tokyo, 1985.

[25] A. Unterberger, *Oscillateur harmonique et opérateurs pseudodifférentiels*, Ann. Inst. Fourier Grenoble **29**,3 (1979), 201–221.

[26] A. Unterberger, J. Unterberger, *La série discrète de $SL(2,\mathbb{R})$ et les opérateurs pseudo-différentiels sur une demi-droite*, Ann. Sci. Ecole Norm. Sup. **17** (1984), 83–116.

[27] A. Unterberger, *Symbolic calculi and the duality of homogeneous spaces*, Contemp. Math. **27** (1984), 237–252.

[28] A. Unterberger, *L'opérateur de Laplace-Beltrami du demi-plan et les quantifications linéaire et projective de $SL(2,\mathbb{R})$* in *Colloque en l'honneur de L. Schwartz*, Astérisque **131** (1985), 255–275.

[29] A. Unterberger, J. Unterberger, *Quantification et analyse pseudodifférentielle*, Ann. Sci. Ecole Norm. Sup. **21** (1988), 133–158.

[30] A. Unterberger, *Quantization and non-holomorphic modular forms*, Lecture Notes in Math. **1742**, Springer-Verlag, Berlin-Heidelberg, 2000.

[31] A. Unterberger, *Automorphic pseudodifferential analysis and higher-level Weyl calculi*, Progress in Math., Birkhäuser, Basel-Boston-Berlin, 2002.

[32] A. Unterberger, *A spectral analysis of automorphic distributions and Poisson formulas*, Ann. Inst. Fourier **54**(5) (2004), 1151–1196.

[33] A. Unterberger, H. Upmeier, *The Berezin Transform and Invariant Differential Operators*, Comm. Math. Phys. **164** (1994), 563–597.

[34] G. Voronoï, *Sur le développement, à l'aide des fonctions cylindriques, des sommes doubles* $\sum f(pm^2 + 2qmn + rn^2)$, Ver. Math. Kongr. Heidelberg (1904), 241–245.

[35] A. Weil, *Sur certains groupes d'opérateurs unitaires*, Acta Math. **111** (1964), 143–211.

[36] H. Weyl, *Gruppentheorie und Quantenmechanik*, reprint of 2nd edition, Wissenschaftliche Buchgesellschaft, Darmstadt, 1977.

[37] D. Zagier, *Introduction to modular forms*, in *From Number Theory to Physics* (M. Waldschmidt, P. Moussa, J.M. Luck and C. Itzykson, eds.), Springer-Verlag, Berlin (1992), 238–291.

Operator Theory: Advances and Applications (OT)

Edited by
Israel Gohberg, Tel Aviv University, Israel

This series is devoted to the publication of current research in operator theory, with particular emphasis on applications to classical analysis and the theory of integral equations, as well as to numerical analysis, mathematical physics and mathematical methods in electrical enginee-ring.

Progress in Mathematics (PM)

Edited by
Hyman Bass, University of Michigan, USA
Joseph Oesterlé, Institut Henri Poincaré, Université Paris VI, France
Alan Weinstein, University of California, Berkeley, USA

Progress in Mathematics is a series of books intended for professional mathematicians and scientists, encompassing all areas of pure mathematics. This distinguished series, which began in 1979, includes research level monographs, polished notes arising from seminars or lecture series, graduate level textbooks, and proceedings of focused and refereed conferences. It is designed as a vehicle for reporting ongoing research as well as expositions of particular subject areas.

BIRKHÄUSER